PCR
Second Edition

Michael J. McPherson

Institute of Molecular and Cellular Biology
Faculty of Biological Sciences
University of Leeds, Leeds, UK

and

Simon Geir Møller

Department of Mathematics and Natural Sciences
Faculty of Science and Technology
University of Stavanger, Stavanger, Norway

Taylor & Francis
Taylor & Francis Group

Published by:
Taylor & Francis Group

In US: 270 Madison Avenue
           New York, N Y 10016
In UK: 4 Park Square, Milton Park
           Abingdon, OX14 4RN

© 2006 by Taylor & Francis Group

First published 2000; Second edition published 2006

ISBN: 0-4153-5547-8

A catalog record for this book is available from the British Library.

Library of Congress Cataloging-in-Publication data has been applied for.

Editor: Elizabeth Owen
Editorial Assistant: Kirsty Lyons
Production Editor: Karin Henderson
Typeset by: Phoenix Photosetting, Chatham, Kent,UK
Printed by: MPG BOOKS Limited, Bodmin, Cornwall, UK

Printed on acid-free paper

10 9 8 7 6 5 4 3 2 1

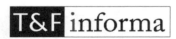

Taylor & Francis Group
is the Academic Division of Informa UK Limited

Visit our web site at http://www.garlandscience.com

Contents

Abbreviations

8-MOP	8-methoxypsoralen
8-oxo-dGTP	8-oxo-2'deoxyguanosine
AFLP	amplified length polymorphism
AMV	avian myeloblastoma virus
AP	alkaline phosphatase
AP-PCR	arbitrarily primed PCR
ARMS	amplification refractory mutation system
ASA	allele specific amplification
ASP	allele-specific PCR
BAC	bacterial artificial chromosome
BCIP	5-bromo, 4-chloro, 3-indolyl phosphate
CAPS	cleaved amplified polymorphic sequence analysis
*Ccd*B	control of cell death
Ct	threshold cycle
CCD	charge coupled device
cDNA	complementary DNA
C_T	comparative threshold
DHPLC	denaturing-high-performance liquid chromatography
DIG	digoxigenin
DIG-dUTP	digoxigenin-11-2'-deoxyuridine-5'-triphosphate
DOP-PCR	degenerate oligonucleotide primed-PCR
dPTP	6-(2-deoxy-β-D-ribofuranosyl)-3,4-dihydro-8H-pyrimido-[4,5-C][1,2]oxazin-7-one
ELISA	enzyme linked immunosorbent assay
EST	expressed sequence tag
FAM	6-carboxyfluorescein
FDD	fluorescent differential display
FRET	fluorescence resonance energy transfer
FS	fluorescent sequencing
GAPDH	glyceraldehyde-3-phosphate dehydrogenase
GAWTS	gene amplification with transcript sequencing
GM	genetically modified
HEX	4,7,2',4',5',7'-hexachloro-6-carboxyfluorescein
HRP	horseradish peroxidase
IPCR	inverse polymerase chain reaction
LCR	ligase chain reaction
LIC	ligation-independent cloning
M-MLV	Moloney murine leukemia virus
MPSV	mutations, polymorphisms and sequence variants
mrPCR	multiplex restriction site PCR
MVR	minisatellite variant repeat
NBT	nitro blue tetrazolium
NF	nonfluorescent
nt	nucleotides
ORFs	open reading frames
PAGE	polyacrylamide gel electrophoresis
PASA	PCR amplification of specific alleles
PBS	phosphate buffered saline
PCR	polymerase chain reaction
PCR-VNTRs	PCR highly polymorphic variable number tandem repeats
PEETA	primer extension, electrophoresis, elution, tailing, amplification
PMBC	peripheral blood mononuclear cells
PMT	photomultiplier tube

PNA	peptide nucleic acid	**StEP**	staggered extension process
PORA	NADPH: protochlorophyllide oxidoreductase	**STR**	short tandem repeats
RACE	rapid amplification of cDNA ends	**TAIL-PCR**	thermal asymmetric interlaced PCR
RACHITT	random chimeragenesis on transient templates	**TAMRA**	6-carboxytetramethyl-rhodamine
RAPD	random amplified polymorphic DNA	**TBR**	tris (2,2′-bipyridine) ruthenium (II) chelate
RAWIT	RNA amplification with *in vitro* translation	**TCA**	trichloroacetic acid
RAWTS	RNA amplification with transcript sequencing	**TdT**	terminal deoxynucleotidyl-transferase
RFLP	restriction fragment length polymorphism	**TEMED**	N,N,N′,N′-tetramethylenediamine
RISC	RNA-induced silencing complex	**TET**	4,7,2′,7-tetrachloro-6-carboxy fluorescein
RNAi	RNA interference	**TK**	thymidine kinase
RT	reverse transcriptase	**Tm**	melting temperature
SDS	sodium dodecyl sulfate	**TNF**	tumor necrosis factor
siRNAs	small interfering RNAs	**TOPO ligation**	topoisomerase-mediated ligation
SNPs	single nucleotide polymorphisms	**Tp**	optimized annealing temperature
SOEing	splicing by overlap extension	**UNG**	uracil *N*-glycosylase
SPA	scintillation proximity assay	**USE**	unique site elimination
SSCP	single strand conformation polymorphism analysis	**VNTR**	variable number tandem repeats
		YAC	yeast artificial chromosome

Preface

The concept underlying this book has not changed from the first edition; it is to provide an introductory text that is hopefully useful to undergraduate students, graduate students and other scientists who want to understand and use PCR for experimental purposes. Although applications of PCR are provided these do not represent a comprehensive catalogue of all possible PCR applications, but serve to indicate the types of application possible. The main purpose of this new edition of *PCR*, as for the first edition, is to provide information on the fundamental principles of the reactions occurring in a PCR tube. Understanding these basic features is essential to fully capitalize upon and adapt the power of PCR for a specific application. This means that the structure of the book remains similar to that of the first edition. The first six chapters discuss the fundamental aspects of performing PCR and of analyzing and cloning the products. All these chapters have been updated and additional aspects added where appropriate. In some Sections there is discussion of particular enzymes or instruments. However, clearly suppliers are continually changing their formulations or designs and so these are provided only to indicate the different types. We recommend checking manufacturers' literature for new and improved systems, particularly when it comes to investing in the purchase of a new PCR instrument. In terms of the applications, a new chapter has been written on real-time PCR, which represents a very sensitive and reliable method for providing information about the relative concentrations of starting template molecules, such as mRNA or genomic genes. The remaining chapters have been updated and protocols have been rationalized to retain those that are likely to be the most useful. We have also removed the list of web addresses of various reagent suppliers. Such lists can quickly become outdated and it is simpler for the reader to identify the up to date website from a web search engine. We hope that this book will provide the basic information required to get scientists started with PCR experiments either to use it simply as a routine tool, or as a starting point for developing new and innovative processes.

We thank those who kindly provided figures to illustrate aspects of the book, and Liz Owen at Garland Science, Taylor & Francis Group for her persistence in ensuring that we kept working on this volume and finished at least close to one of the deadlines!

An introduction to PCR

1.1 Introduction: PCR, a 'DNA photocopier'

Does it really work? It is so simple! Why did I not think of it? These thoughts were probably typical of most molecular biologists on reading early reports of the polymerase chain reaction or PCR as it is more commonly called. PCR uses a few basic everyday molecular biology reagents to make large numbers of copies of a specific DNA fragment in a test-tube. PCR has been called a 'DNA photocopier'. While the concept is simple, PCR is a complicated process with many reactants. The concentration of template DNA is initially very low but its concentration increases dramatically as the reaction proceeds and the product molecules become new templates. Other reactants, such as dNTPs and primers, are at concentrations that hardly change during the reaction, while some reactants, such as DNA polymerase, can become limiting. There are significant changes in temperature and pH and therefore dramatic fluctuations in the dynamics of a range of molecular interactions. So, PCR is really a very complex process, but one with tremendous power and versatility for DNA manipulation and analysis.

In the relatively short time since its invention by Kary Mullis, PCR has revolutionized our approach to molecular biology. The impact of PCR on biological and medical research has been like a supercharger in an engine, dramatically speeding the rate of progress of the study of genes and genomes. Using PCR we can now isolate essentially any gene from any organism. It has become a cornerstone of genome sequencing projects, used both for determining DNA sequence data and for the subsequent study of putative genes and their products by high throughput screening methodologies. Having isolated a target gene we can use PCR to tailor its sequence to allow cloning or mutagenesis or we can establish diagnostic tests to detect mutant forms of the gene. PCR has become a routine laboratory technique whose apparent simplicity and ease of use has allowed nonmolecular biology labs to access the power of molecular biology. There are many scientific papers describing new applications or new methods of PCR. Many commercial products and kits have been launched for PCR applications in research and for PCR-based diagnostics and some of these will be discussed in later chapters.

1.2 PCR involves DNA synthesis

PCR copies DNA in the test-tube and uses the basic elements of the natural DNA synthesis and replication processes. In a living cell a highly complex system involving many different proteins is necessary to replicate the complete genome. In simplistic terms, the DNA is unwound and each strand of the parent molecule is used as a template to produce a complementary 'daughter' strand. This copying relies on the ability of nucleotides to base pair according to the well-known Watson and Crick rules; A always pairs with T and G always pairs with C. The template strand therefore specifies the base sequence of the new complementary DNA strand. A large number of proteins and other molecules, such as RNA primers, are required to ensure that the process of DNA replication occurs efficiently with high fidelity, which means with few mistakes, and in a tightly regulated manner. DNA synthesis by a DNA polymerase must be 'primed', meaning we need to supply a short DNA sequence called a primer that is complementary to a template sequence. Primers are synthetically produced DNA sequences usually around 20 nucleotides long. The DNA polymerase will add nucleotides to the free 3'-OH of this primer according to the normal base pairing rules (*Figure 1.1*).

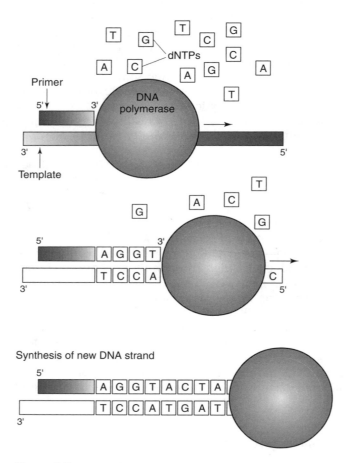

Figure 1.1

Primer extension by a DNA polymerase. The primer anneals to a complementary sequence on the template strand and the DNA polymerase uses the template sequence to extend the primer by incorporation of the correct deoxynucleotide (dNTP) according to base pairing rules.

PCR requires only some of the components of the complex replication machinery to copy short fragments of DNA in a simple buffer system in a test tube. Unwinding of the DNA in the cell uses a multi-component complex involving a variety of enzymes and proteins, but in PCR this is replaced simply by a heating step to break the hydrogen bonds between the base pairs of the DNA duplex, a process called denaturation.

Following template denaturation two sequence-specific oligonucleotide primers bind to their complementary sequences on the template DNA strands according to normal base pairing rules (*Figure 1.2*). These primers define the region of template to be copied. DNA polymerase then begins to add deoxynucleotides to the 3'-OH group of both primers producing new duplex DNA molecules (*Figure 1.2*). This requirement of DNA polymerases to use primers to initiate DNA synthesis is critical for the PCR process since it means we can control where the primers bind, and therefore which region of DNA will be replicated and amplified. If the DNA polymerase was like an RNA polymerase that does not require a primer then we would have no way of defining what segment of DNA we wanted to be copied.

At the next heating step the double-stranded molecules, which are heteroduplexes containing an original template DNA strand and a newly synthesized DNA strand produced during the first DNA synthesis reaction, are now denatured. Each DNA single strand can now act as a template for the next round of DNA synthesis. As discussed in detail in Chapter 2, it is during this second cycle of PCR that the first DNA single strand of a length defined by the positions of the primers can be formed. In cycle 3 the first correct length double-stranded PCR products are formed. In subsequent cycles there is then an exponential increase in the number of copies of the 'target' DNA sequence; theoretically, the number of copies of the target sequence will be doubled at each PCR cycle. This means that at 100% efficiency, each template present at the start of the reaction would give rise to 10^6 new strands after only 20 cycles of PCR. Of course the process is not 100% efficient, and it is usually necessary to carry out more reaction cycles, often 25 to 40 depending upon the concentration of the initial template DNA, its purity, the precise conditions and the application for which you require the product. The specificity and efficiency of PCR, however, means that very low numbers of template molecules present at the start of the PCR can be amplified into a large amount of product DNA, often a microgram or more, which is plenty for a range of detailed analyses. Of course, this ability to amplify also means that if you happen to contaminate your reaction with a few molecules of product DNA from a previous reaction, you may get a false result. This is why performing control reactions is so important and we will deal with such contamination problems in Chapter 4.

1.3   PCR is controlled by heating and cooling

PCR relies on the use of different temperatures for the three steps of the reaction, denaturation, annealing and extension. A high temperature, usually 94–95°C, is used to denature (separate) the strands of the DNA template. The temperature is then lowered to allow the primers to anneal by base pairing to their complementary sequences on the template strands; this temperature varies depending on the primers (see details in Chapter 3).

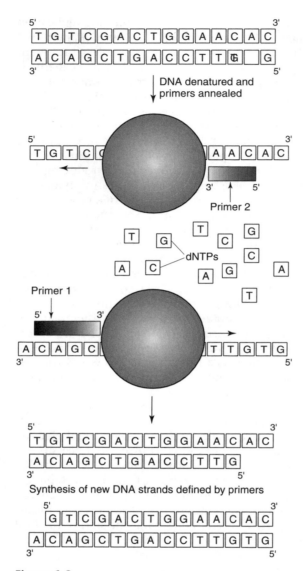

Figure 1.2

The first cycle of a PCR. A double-stranded template molecule is denatured. Primers anneal to their complementary sequences on the single-stranded template. DNA synthesis is catalyzed by a thermostable DNA polymerase. The result of this PCR cycle is that two copies of the target sequence have been generated for each original copy.

The annealing temperature is important to ensure high specificity in the reaction; generally the higher the annealing temperature the more specific will be the reaction. A temperature of 55°C is commonly used, but in many cases a higher temperature is better and this can even be as high as 72°C for some experiments, leading to a two-temperature PCR cycle. Finally, for efficient DNA synthesis, the temperature is adjusted to be optimal for the DNA polymerase activity, normally 72°C (see Chapter 3). To amplify the target DNA it is necessary to cycle through these temperatures several times (25 to 40 depending on the application). Conveniently, this temperature cycling is accomplished by using a thermal cycler, a programmable instrument that can rapidly alter temperature and hold samples at the desired temperature for a set time. This automation is one of the important advances that led to PCR becoming widely accessible to many scientists and is covered in more detail in Chapter 3. Before thermal cyclers became available, PCR was performed by using three water baths set to temperatures of typically 95°C, 55°C and 72°C, and reaction tubes in racks were moved manually between the baths.

The other major technological advance that preceded the development of thermal cyclers was the replacement of DNA polymerase I Klenow fragment with thermostable DNA polymerases, such as *Taq* DNA polymerase, which are not inactivated at the high denaturation temperatures used during PCR. The ability to carry out the reaction at high temperatures enhances the **specificity** of the reaction (Chapter 4). At 37°C, where Klenow works best, primers can bind to nontarget sequences with weak sequence similarity, because mismatches between the two strands can be tolerated. This leads to poor specificity of primer annealing and the amplification of many nontarget products. The introduction of thermostable DNA polymerases also reduced the cost of a reaction by reducing the amount of polymerase required. With Klenow, at each denaturing step the enzyme was also denatured and therefore a fresh aliquot had to be added at each cycle. Thermostable polymerases retain their activity at the denaturation temperatures and therefore only need to be added at the start of the reaction.

1.4 PCR applications and gene cloning

PCR has revolutionized our approach to basic scientific and medical research, to medical, forensic and environmental testing. It provides an extremely flexible tool for the research scientist, and every molecular biology research laboratory now uses PCR routinely; often adapting and tailoring the basic procedures to meet their own special needs. It has become an indispensable tool for routine and repetitive DNA analyses such as diagnosis of certain genetic diseases within clinical screening laboratories where speed and accuracy are important factors, and also for sample identification in forensic and environmental testing. In particular PCR has become a central tool in the analysis and exploitation of genome sequence information, for example in gene knockout through RNA interference where PCR allows the rapid generation of appropriate constructs. It also facilitates measurement of levels of gene expression by 'real-time' PCR that monitors the level of product amplification at each cycle of the PCR (Chapter 9), providing information on the relative concentrations of template cDNA.

In some cases PCR provides an alternative to gene cloning, but in other cases it provides a complementary tool. In gene cloning a fragment of DNA is joined by ligation to a cloning vector which is able to replicate within a host cell such as the bacterium *Escherichia coli*. As the bacterium grows, the new recombinant DNA molecule is copied by DNA replication, and as the cell divides the number of cells carrying the recombinant molecule increases. Finally, when there are enough cells you can isolate the recombinant DNA molecules to provide sufficient DNA for analysis or further manipulation of the cloned DNA fragment. This type of cloning experiment takes about 2–3 days. PCR also amplifies your target DNA fragment so that you have enough to analyze or manipulate, but in this case the DNA replication occurs in a test-tube and usually takes no more than 1–3 hours. In many cases, for example in diagnostic tests for cancers or genetic diseases, including ante-natal screening, or in forensic testing, PCR provides the most sensitive and appropriate approach to analyze DNA within a day.

For studying new genes and genetic diseases it is often necessary to create gene libraries and this may involve PCR followed by cloning into a suitable vector. Also many experiments to study structure and function require expression in host cells and this requires the cloning of the gene protein perhaps as a PCR product, into a suitable expression vector. So in many cases PCR and gene cloning represent complementary techniques. It is important to consider carefully the most appropriate strategy for the experiments you wish to undertake. Integrating PCR and cloning will be covered further in Chapter 6 while diagnostic applications of PCR are covered in Chapter 11.

1.5 History of PCR

As long ago as 1971, Khorana and colleagues described an approach for replicating a region of duplex DNA by using two DNA synthesis primers designed so that their 3'-ends pointed towards each other (1). However, the concept of using such an approach repeatedly in an amplification format was not conceived for another 12 years. 'Sometimes a good idea comes to you when you are not looking for it.' With these words, Kary Mullis, the inventor of PCR, starts an account in *Scientific American* of how, during a night drive through the mountains of Northern California in Spring 1983, he had a revelation that led him to develop PCR (Mullis, 1990). Mullis was awarded the 1993 Nobel Prize for Chemistry for his achievement. The practical aspects of the PCR process were then developed by scientists at Cetus Corporation, the company for which Mullis worked at that time. They demonstrated the feasibility of the concept that Mullis had provided, and PCR became a major part of the business of Cetus, before they finally sold the rights to PCR in 1991 for $300m to Roche Molecular Systems. PCR and the thermostable polymerase responsible for the process were named as the first 'Molecule of the Year' in 1989 by the international journal *Science*.

Since the myriad of applications of PCR were recognized it has become rather entangled in commercialism, due to the large amounts of money to be made from licensing the technology. PCR is covered by patents, granted to Hoffman La-Roche and Roche Molecular Systems, and these have been vigorously enforced to prevent unlicensed use of the method. Some of these patents terminated on 28 March 2005. From this date it has been possible to perform basic PCR in the US without a license, although some other patents still apply to instruments and specific applications. Outside the US in countries covered by the equivalent patents, there is a further year of patent protection to run.

Key milestones in the development of PCR

1983   Kary Mullis of Cetus Corp. invents PCR.

1985   First paper describing PCR using Klenow fragment of DNA polymerase I (2).

1986   Cetus Corp. and Perkin Elmer Corp. establish a joint venture company (Perkin Elmer Cetus) to develop both instruments and reagents for the biotechnology research market.

1987   Cetus develop a partnership with Kodak for PCR-based diagnostics, but Kodak terminate this agreement and Hoffman-La Roche become the new partner.

1988   First paper describing the use of *Taq* DNA polymerase in PCR (3).

1990   Cetus licence certain reagents companies, namely Promega, Stratagene, USB, Pharmacia, Gibco-BRL and Boehringer, to sell native *Taq* DNA polymerase for non–PCR applications.

1991   Cetus wins court case against DuPont who challenged the Cetus PCR patents.

1991   Perkin Elmer Cetus joint venture dissolved as the PCR rights are acquired by Hoffman LaRoche.

1991   Perkin Elmer form a 'strategic alliance' with Roche to sell PCR reagents in the **research** market. Roche continue to develop the **diagnostics** reagents business. Perkin Elmer assume total responsibility for the thermal cycler business.

1991   Cetus is acquired by Chiron Corp. for non-PCR business aspects, in particular interleukin-2-based pharmaceuticals.

1993   Roche file a lawsuit against Promega for alleged infringement of their license to sell native *Taq* polymerase for non-PCR applications. Action also taken against several smaller companies for selling *Taq* DNA polymerase without license agreements. These disputes between Roche and Promega are still proceeding through the courts in 2005, and do not look like they will be resolved quickly or easily.

1993   Perkin Elmer merges with Applied Biosystems. The Applied Biosystems Division of Perkin Elmer assumes responsibility for all DNA products such as DNA synthesis, sequencing and the PCR in addition to the other products associated with protein sequencing and analysis.

1993   License granted to Boehringer-Mannheim to supply reagents, including *Taq* polymerase for use in PCR.

1993   Kary Mullis, inventor of PCR, wins a Nobel Prize for Chemistry.

1993 + Widespread licensing of PCR technology and *Taq* DNA polymerases to a large number of Biological Supplies Companies.

1998   Promega Corporation challenges the original patents on native *Taq* DNA polymerase and court proceedings continue.

2005   March 28 2005 is the date on which several US PCR patents expire:
   - 4 683 195 Process for amplifying, detecting and/or cloning nucleic acid sequences;

- 4 683 202 Process for amplifying nucleic acid sequences;
- 4 965 188 Process for amplifying, detecting and/or cloning nucleic acid sequences using a thermostable enzyme;
- 6 040 166 Kits for amplifying and detecting nucleic acid sequences including a probe;
- 6 197 563 Kits for amplifying and detecting nucleic acid sequences;
- 4 800 159 Process for amplifying, detecting and/or cloning nucleic acid sequences;
- 5 008 182 Detection of AIDS-associated virus by PCR;
- 5 176 995 Detection of viruses by amplification and hybridization.

The speed and simplicity of PCR technology accompanied by an increased range of high quality products has led to a more rational approach to PCR experimentation. We understand better the molecular processes underlying PCR (see Chapter 2) so that it is seen less as a 'witches' brew'. It is important to highlight good practices that increase confidence in results by reducing the likelihood of artefactual results. The importance of good PCR technique, particularly with regard to proper controls and the prevention of contamination (Chapter 4) cannot be overemphasized. Remember, if you work in a research laboratory a wrong result may be inconvenient leading to a waste of time, effort and money and so should be avoided. But, if you work in a diagnostic laboratory, a wrong result could mean the difference between life and death. It is a good idea to start with the highest standards and expectations so that you can be confident in your results no matter where you work.

PCR has now been adapted to serve a variety of applications and some of these will be described in this book (Chapters 5 to 11).

Further reading

Mullis KB (1990) The unusual origins of the polymerase chain reaction. *Sci Am* **262**: 56–65.

White TJ (1996) The future of PCR technology: diversification of technologies and applications. *Trends Biotechnol* **14**: 478–483.

References

1. Kleppe K, Ohtsuka E, Kleppe R, Molineux R, Khorana HG (1971) Studies on polynucleotides. XCVI. Repair replication of short synthetic DNA's as catalysed by DNA polymerases. *J Mol Biol* **56**: 341–346.
2. Saiki RK, Scharf S, Faloona F, Mullis KB, Horn GT, Erlich HA, Arnheim N (1985) Enzymatic amplification of beta-globin genomic sequences and restriction site analysis for diagnosis of sickle cell anemia. *Science* **230**: 1350–1354.
3. Saiki RK, Gelfand DH, Stoffel S, Scharf S, Higuchi R, Horn GT, Mullis KB, Erlich HA (1988) Primer-directed enzymatic amplification of DNA with a thermostable DNA polymerase. *Science* **239**: 487–491.

Understanding PCR

2

This Chapter is designed to provide you with essential information to understand what is happening in the PCR tube. We will consider the kinetics of the PCR process during the various stages of the reaction and then outline a basic protocol as a starting point for many PCR experiments.

2.1   How does PCR work?

PCR proceeds in three distinct steps governed by temperature.

- Denaturation: the double-stranded template DNA is denatured by heating, typically to 94°C, to separate the complementary single strands.
- Annealing: the reaction is rapidly cooled to an annealing temperature to allow the oligonucleotide primers to hybridize to the template. The single strands of the template are too long and complex to be able to reanneal during this rapid cooling phase. During this annealing step the thermostable DNA polymerase will be active to some extent and will begin to extend the primers as soon as they anneal to the template. This can lead to specificity problems if the annealing temperature is too low (Chapter 4).
- DNA synthesis: the reaction is heated to a temperature, typically 72°C for efficient DNA synthesis by the thermostable DNA polymerase.

In the first cycle of PCR each template strand gives rise to a new duplex, as shown in *Figure 2.1(A)*, doubling the number of copies of the target region. Likewise at each subsequent cycle of denaturation, annealing and extension, there is a theoretical doubling of the number of copies of the target DNA. If PCR achieved 100% efficiency then 20 cycles would yield a one million-fold amplification of the target DNA (2^{20} = 1 048 572). Of course PCR is not 100% efficient for a variety of reasons that we will consider shortly, but by increasing the number of cycles and optimizing conditions amplification by 10^6-fold or greater is routinely achievable.

One of the great advantages of PCR is its ability to amplify a defined region of DNA from a very complex starting template such as genomic DNA. It is therefore worth dissecting what is happening during PCR amplification from a genomic DNA template as this will provide a better understanding of the reaction process (Section 2.3).

PCR uses two oligonucleotide primers that act as sites for initiation of DNA synthesis by the DNA polymerase and so these primers define the region of the template DNA that will be copied (1). DNA polymerases need a primer to begin DNA synthesis and so we need to know at least small parts of the DNA sequence of the target region in order to be able to design these primers. The primers (sometimes called amplimers) are complementary to regions of known sequence on opposite strands of the template DNA and their 3'-OH end points towards the other primer. The primer is

(A) The first cycle of a PCR reaction

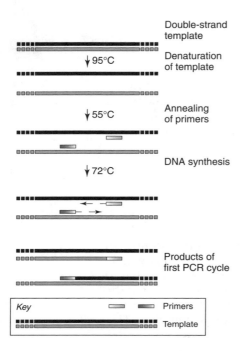

Double-strand template

Denaturation of template

Annealing of primers

DNA synthesis

Products of first PCR cycle

Key — Primers
— Template

(B) The second cycle of a PCR reaction

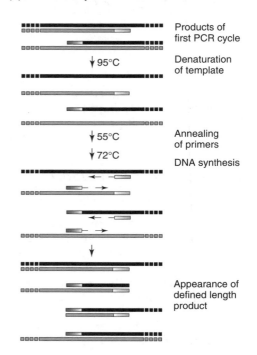

Products of first PCR cycle

Denaturation of template

Annealing of primers

DNA synthesis

Appearance of defined length product

(C) The third cycle of a PCR reaction

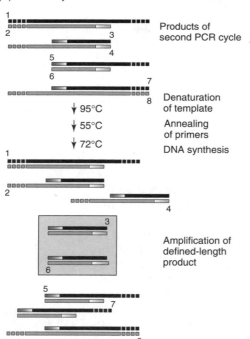

Products of second PCR cycle

Denaturation of template

Annealing of primers

DNA synthesis

Amplification of defined-length product

(D) The fourth cycle of a PCR reaction

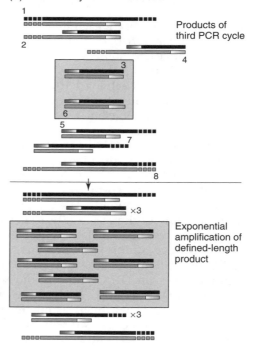

Products of third PCR cycle

Exponential amplification of defined-length product

extended by the DNA polymerase incorporating the four deoxynucleotides (dATP, dGTP, dCTP and dTTP) in a template-directed manner. The DNA sequence between the two primer binding sites will therefore be replicated during each cycle of the PCR. The reaction vessel, a 0.2 ml or 0.5 ml polypropylene microcentrifuge tube or well of a microtiter plate, is placed in a thermal cycler and subjected to a series of heating and cooling reactions as outlined in *Figure 2.2*.

A typical PCR protocol is provided at the end of this Chapter in *Protocol 2.1* and so you should refer to this as issues are highlighted in the remainder of this Chapter.

At the start of a PCR there is usually an extended denaturation step at 94°C for 2–5 min to ensure that the template DNA is efficiently denatured. There are then usually three temperature-controlled steps:

- 94°C to denature the template strands; then
- 40–72°C (55°C is often used as a good starting point) to allow the primers to anneal; then
- 72°C, the optimal temperature for many thermostable DNA polymerases to allow efficient DNA synthesis (2).

These three steps are repeated usually for between 25 and 40 times, as necessary, for the specific application. Normally there is then an extended 72°C step to ensure that all of the products are full-length. Finally the reaction is cooled to either room temperature or 4°C depending upon the application and type of thermal cycler used.

2.2   PCR: a molecular perspective

A good way to understand any molecular biology process is to think about what is going on at the molecular level. Try to imagine what is happening to the different types of molecules in a reaction tube. Ask yourself questions about the reactants and what will happen to these as the reaction proceeds.

- What are the relative concentrations of the various reactants?
- Which reactants are present in excess and which are limiting?
- What interactions are going on between molecules such as enzymes and DNA?
- What factors will influence these molecular interactions?
- What are the activities of the enzyme and how will these modify the DNA?
- What are the products of the reaction and how will their accumulation affect the reaction?

Figure 2.1 (opposite)

PCR theoretically doubles the amount of target DNA at each cycle. (A) Cycle 1, products generated from template DNA are not of a defined length. (B) Cycle 2, the first single-strand products of defined length are produced due to priming on single-strand products generated during cycle 1. (C) Cycle 3 results in the production of the first double-strand products of defined length. (D) Cycle 4 and subsequent cycles lead to exponential amplification of the defined length products. In parts C and D the various strands are numbered to enable the templates and products to be followed.

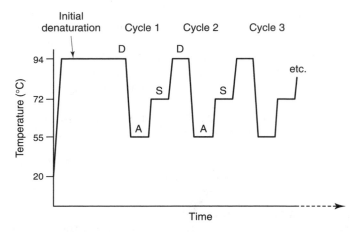

Figure 2.2

Representation of thermal cycling during a PCR. The reaction is heated from room temperature to an initial denaturation phase of around 5 min at 94°C to ensure the original template strands are now single-stranded. There then follows a series of repeated cycling steps through temperatures for denaturation of double-stranded molecules (D), annealing of primers to template (A) and DNA synthesis from the primer (S).

It is sometimes useful to think about a single enzyme molecule in the reaction tube and to consider how it works to gain a molecular perspective on the reaction.

A genomic DNA template

PCRs are usually performed on template DNA molecules that are longer than the target region that we wish to amplify. The extreme case is where we start with genomic DNA. A key question is 'How does the DNA polymerase know when it has reached the end of the target region that is to be copied?' The answer is that it does not know; it therefore carries on synthesizing new DNA until the temperature of the reaction is increased during the denaturing step of the next PCR cycle (see *Figure 2.1(A)*). If we think about a simple case where we start with one molecule of genomic DNA, then, after one cycle of PCR we will have the original template strands and two new strands, initiated from the primers. These new strands will be much shorter than the original genomic strands, but will still be longer than the target region to be amplified. Importantly however, one end of each of the new strands now corresponds to a primer sequence. In the second cycle, the primers again anneal to the original templates but also to the strands synthesized during the first cycle. The DNA polymerase will extend from the primers, and again the original templates will give rise to longer strands of undefined length. However, on the strands synthesized during the first cycle the enzyme will 'run out' of template DNA when it reaches the end of the primer sequence incorporated during the first cycle. So, by the end of this second cycle we have produced two single strands of

DNA that correspond to the product length defined by the two primers (*Figure 2.1(B)*). These defined-length strands are now amplified in each subsequent cycle leading to an exponential accumulation of this target PCR product. This is illustrated in *Figure 2.1(C) and (D)* and *Table 2.1*.

This exponential amplification of the target PCR product contrasts dramatically with the linear accumulation of the longer strands copied from the original template molecule. Every PCR cycle produces only two further elongated DNA strands for each original template DNA molecule. As you can also see from *Table 2.1* by the end of 20 cycles in an 'ideal PCR', for every original template molecule there will only be 42 single strands of DNA of undefined length, including the two original template strands. So the theoretical 10^6 double-strand product molecules of correct length generated for each original template duplex are present in vast excess over these strands of undefined length.

As illustrated in *Table 2.1*, amplification at 100% efficiency should generate some 10^6 product molecules per original template molecule. So, under these ideal conditions starting with 1 µg of human genomic DNA (around 3×10^5 molecules) a single copy target sequence should theoretically be amplified to yield 3×10^{11} product fragments after 20 cycles. In practice, as with most biological reactions, PCR amplification is not 100% efficient, so normally a greater number of cycles (25–40) are performed to achieve these levels of amplification.

Table 2.1 Theoretical accumulation of PCR products during the first 20 cycles of a PCR with a single genomic DNA template

Cycle number	Number of single strands of undefined length	Number of single strands of defined length	Number of copies of double-strand target
0	2	0	1 [a]
1	4	0	2
2	6	2	4
3	8	8	8
4	10	22	16
5	12	52	32
6	14	114	64
7	16	240	128
8	18	494	256
9	20	1 004	512
10	22	2 026	1 024
11	24	4 072	2 048
12	26	8 164	4 096
13	28	16 356	8 192
14	30	32 738	16 384
15	32	65 504	32 768
16	34	131 038	65 536
17	36	262 108	131 072
18	38	524 250	262 144
19	40	1 048 536	524 288
20	42	2 097 110	1 048 576

[a] This copy represents the original target DNA which therefore represents two single strands of undefined length.

Table 2.2 Concentrations of reactants and products before and after a 30-cycle PCR. Some components undergo dramatic alterations in concentration while others show little change

Reagent	Initial reaction conditions				Conditions following 30 cycles of PCR (10^6-fold amplification)				
	Amount	Picomoles	Concentration	Ratio to template	Amount	Picomoles	Concentration	Ratio to genomic template	Ratio to amplified fragment
Human genomic DNA	1 μg	5×10^{-7}	5 fM	1	1 μg	5×10^{-7}	5 fM	1	10^{-6}
Target region (1 kb)	0.3 pg	5×10^{-7}	5 fM	1	0.3 μg	0.5	5 nM	10^6	1
Each primer	325 ng	50	0.5 μM	10^8	322 ng	49.5	0.495 μM	10^8	99
Each dNTP	2.88 μg	5×10^3	50 μM	10^{10}	2.78 μg	4.8×10^3	48 μM	9.5×10^9	9.5×10^3
Taq DNA polymerase	2 units	0.1	1 nM	2×10^5	2 units	0.1	1 nM	2×10^5	0.2

Table 2.2 illustrates the relative concentrations and numbers of molecules present in an ideal PCR starting with a human genomic DNA template. The numbers at the start of the reaction and then after 20 cycles of amplification are shown. It is clear that some reactant and product concentrations change substantially whilst others do not. It will be useful to refer to *Table 2.2* during some of the following discussion.

2.3   The kinetics of PCR

We can consider a PCR to have three distinct phases as shown in *Figure 2.3*:

- **E**: the early cycles during which the primers search the template DNA for their complementary sequences, effectively acting like probes in a DNA hybridization experiment;
- **M**: the mid cycles when the amplification process is well underway with primer pairs acting together to bring about an exponential accumulation of the product fragment; and
- **L**: the late cycles, sometimes called the plateau, when amplification is suboptimal due to limiting reagents (most usually the thermostable DNA polymerase) or inhibition of the reaction.

Ideally we want to enhance the specificity of primer selection during **E**, achieve maximal efficiency of amplification during **M** and stop the reaction before **L**. Each phase will now be considered in more detail.

The early cycles (E)

For a PCR from genomic DNA we have relatively few copies of the template and a large number of copies of the two primers that define the target region to be amplified. If this target represents a unique gene, then for each copy of the haploid genome there will only be one specific binding site for each

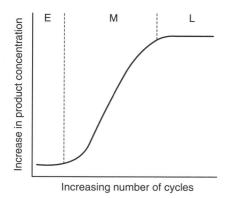

Figure 2.3

Kinetics of accumulation of the target product during PCR. E is the early phase of primer scanning and initial product formation; M is the middle phase during which product accumulates in an exponential manner; L is the late phase or plateau where product accumulation is suboptimal.

primer, a bit like two needles (some 20 nucleotides in length) searching in a haystack (some 6×10^9 nucleotides). A 1 µg aliquot of human genomic DNA contains 3×10^5 copies of the genome and therefore of the single copy target sequence. The number of molecules of each primer is around 1.5×10^{14} so these are present in vast excess over the template. The first task of the primers is to find their complementary sequences; in effect they are acting like hybridization probes scanning the genomic DNA for their complement and since there are plenty of copies of each primer this search process should not be difficult. In this stage the primers will bind transiently to random sequences; if the sequence is not complementary to the primer then it will rapidly dissociate and reanneal elsewhere. The reason primers dissociate from nontarget sequences is because the annealing conditions (temperature, Mg^{2+} ion concentration) favor the formation of perfectly matched duplexes. During this intensive search a primer will find the correct complementary sequence and will remain associated to the template in a binary complex for sufficient time for further interaction with a molecule of thermostable DNA polymerase to form a ternary complex. As the DNA polymerase will display some DNA synthesis activity even at the annealing temperature, which is normally lower than its optimal activity temperature, it will initiate DNA synthesis from this primer. This results in a much more stable complex of template and extended primer that will not dissociate when the temperature is raised to 72°C for optimal DNA polymerase activity. Any primers that are only transiently associated with the DNA will be denatured from the template as the temperature is raised.

Mispriming

During the annealing phase some of the primers may find alternative sites on the template to which they are partially complementary and to which they can bind. Most importantly if their 3′-end is complementary to a random sequence on the template the primer may remain associated for sufficient time for a DNA polymerase molecule to interact with the duplex region and initiate DNA synthesis. This is a mispriming or nonspecific priming event (Chapter 4). The DNA polymerase cannot discriminate between a perfectly matched primer–template duplex and one that has some mismatches. The DNA polymerase acts like a machine; anything that looks like a substrate will be used as a substrate. In many cases such mispriming causes problems during a PCR because the misprimed products, which now have a perfect primer sequence at one end, can interfere with efficient amplification of the true target fragment as described below.

The presence of the primer sequence is obviously critical for successful amplification of the correct sequence during later cycles of the PCR. However, nontarget products generated by mispriming will also have a primer sequence associated with it. Of course, not all of these nonspecific misprimed sequences will be amplified during PCR; amplification can only occur if there is a second priming site for either of the primers, sufficiently close to the first site (within a few kb) and on the opposite strand of the template. Nonetheless, a proportion of nonspecific products often fulfill these criteria and may become amplified, together with, or in preference to, the target sequence.

How can you prevent mispriming? The primers should be designed carefully according to the guidelines outlined in Chapter 3. The annealing temperature should be selected to be as high as possible so that primers can only base pair to their perfectly complementary sequence to form a stable duplex. A variety of approaches have been described to improve the specificity of primer annealing. These include various 'enhancer' additives, 'hot start' and 'touchdown' procedures that are dealt with in Chapter 4. Optimizing the annealing temperature by using a gradient thermocycler is also a useful approach.

Of course there are some cases where conditions are chosen that do allow mismatches between primer and template to be tolerated. For example, in procedures such as PCR mutagenesis (Chapter 7) it is essential that mismatched primers can act as templates and so conditions such as primer length and annealing temperature are adjusted to allow this.

The mid cycles (M) – exponential phase

Following the early cycles of PCR the amplification phase of the reaction begins. The mid phase of a PCR involves the exponential amplification, ideally with a doubling of the number of copies of the target sequence selected during the early cycles. As this phase of PCR gets underway the process of primer scanning for complementary sites becomes simpler as there are an increasing number of copies of the target sequence which contain the primer binding sites. The rapid accumulation of product fragment continues until the efficiency of this amplification is disturbed and the reaction eventually reaches a plateau. It is important to stop a PCR during this exponential phase rather than allowing it to reach the plateau phase.

Late cycles (L) – plateau phase

The plateau phase is reached as a consequence of changes in the relative concentrations of certain components of the reaction (*Table 2.2*). In particular all the molecules of thermostable polymerase (about 3×10^{10}) will be engaged in DNA synthesis. If there are a larger number of product strands than DNA polymerase molecules then not all DNA strands will be used as templates for further DNA synthesis during each cycle and therefore exponential amplification cannot continue. In addition, as product DNA accumulates and the ratio of primer to product decreases, there is a greater tendency for product strands to anneal thus preventing their use as templates. Since the products are longer than the primers the annealing of complementary product strands can begin at higher temperatures than for primer/template annealing therefore product strands can be sequestered from the reaction. It is likely that some nonspecific products will accumulate. As the true product becomes less available to act as template, due to reannealing, any nonspecific products that have been generated will be present at lower concentrations than the true products and therefore can provide alternative templates for amplification. These nontarget products may now accumulate at an exponential rate while the true products will increase in number more slowly. In some cases, at high product concentrations product strands can anneal to allow self-primed concatameric products that are longer than the desired product and can appear as a higher molecular mass smear on an agarose gel. These features are clearly nonproductive and lead to contamination of the true PCR product with other fragments.

In general you are probably best stopping the PCR after 30–35 cycles. If this was insufficient to generate the desired amount of product, use an aliquot of the first PCR as template in a fresh PCR. Of course there are exceptions and some protocols call for more cycles when minute amounts of template are available. For example, in difficult PCR, with a complex template present at low concentration, the accumulation of product fragments will occur more slowly. In such cases a greater number of cycles are needed in order to achieve good amplification before reaching the plateau phase. Some recombination strategies for generating variant libraries of sequences may use up to 60 cycles (Chapter 7) although in general it is better to perform no more than 35 cycles during a PCR, and to reamplify an aliquot. Remember, even when using a proofreading enzyme the greater the number of cycles you perform the greater the risk of mutations being introduced, so DNA sequencing of products or resulting clones is essential.

2.4   Getting started

Protocol 2.1 outlines a basic PCR procedure that provides a good starting point for most applications. You can use any source of template DNA such as genomic DNA, linear or circular plasmid or phage DNA, and more details on template sources are given in Chapter 3, which also considers the various components of the reaction. This basic protocol often gives very good results; in other cases it provides evidence for product and so provides a starting point for optimization experiments as described in Chapter 4. In any PCR it is important that you carry out parallel control experiments as detailed in Chapter 4. As a minimum these should include setting up PCR tubes with all but one reaction component, specifically one without template DNA and one without primers. Other controls could include adding only one of the two primers to check that products are only generated when both primers are present. *Remember to set up control reactions last so that you detect any possible contamination introduced during the set-up of sample tubes.*

2.5   Post-PCR analysis

Once the PCR has finished, you need to analyze the products. The usual way of doing this is to size fractionate the DNA through an agarose gel. Examining the gel provides evidence for success or failure.

- Is there a single product band? Is it of the expected size? This would be a good indication of success, but you should confirm this by further PCR analysis, restriction analysis or DNA sequencing either before or after cloning.
- Are there several products? Is your product the major band? This might indicate suboptimal annealing temperature, but certainly suggests a

problem with the PCR. If the major product is likely to be your product you might isolate this product from the gel and analyze it as above. Alternatively, repeat the PCR by adjusting the conditions to increase stringency.

- Is there a very strong low molecular weight product band? This is usually primer-dimer that results from self-priming of one or both primers to generate a small product that is very efficiently amplified. It may not be a problem if you are still able to see substantial amounts of your product band. If the intensity of your product band is low, then either repeat the reaction at higher stringency, or redesign one of the primers and check for self-complementarity and annealing to the partner primer.

Depending on the success of your first PCR it may be necessary to optimize the conditions to achieve improved results as described in Chapter 4. It will certainly be necessary to confirm the identity of the PCR product to ensure it is the desired sequence to avoid spending time, effort and money studying the wrong DNA fragment.

Later, if you are using a routine procedure optimized for amplification of your product then there are solution approaches that do not require gel analysis or real-time PCR that allow you to follow the kinetics of product accumulation at each cycle of the PCR. The various approaches for PCR analysis are described in Chapter 5.

Further reading

Kidd KK, Ruano G (1995) Optimising PCR. In McPherson MJ, Hames BD, Taylor GR (eds) *PCR2: A Practical Approach*, pp. 1–22. Oxford University Press, Oxford, UK.

References

1. Mullis K, Faloona F (1987) Specific synthesis of DNA *in vitro* via a polymerase-catalyzed chain reaction. *Methods Enzymol* **155**: 335–350.
2. Chien A, Edgar DB, Trela JM (1976) Deoxyribonucleic acid polymerase from the extreme thermophile *Thermus aquaticus*. *J Bacteriol* **127**: 1550–1557.

Protocol 2.1 Basic PCR

EQUIPMENT

Ice bucket

Microcentrifuge

Thermal cycler

Gel electrophoresis tank

MATERIALS AND REAGENTS

Thermostable DNA polymerase and accompanying 10 × reaction buffer[1] (eg. *Taq* DNA polymerase or KOD DNA polymerase)

2 mM dNTP solution

Oligonucleotide primers

Template DNA

Mineral oil

0.8% agarose (100 ml; 0.8 g agarose in 100 ml 1 × TAE)

1. Add the following components to a 0.5 ml microcentrifuge tube:
   (a) 5 μl 10 × PCR buffer (supplied with enzyme);
   (b) 5 μl 2 mM dNTPs;
   (c) 1 μl primer 1 (10 pmol μl⁻¹);
   (d) 1 μl primer 2 (10 pmol μl⁻¹);
   (e) template DNA (~ 0.1 pmol of plasmid to 1 μg genomic DNA);
   (f) thermostable DNA polymerase (1 unit);
   (g) water to 50 μl.
   Ensure that fresh pipette tips are used for each component and make additions to fresh sections of the sides of the tube to prevent mixing of components until all reagents are added[2]. Set up control tubes in the same way but leaving out either DNA or primers.

2. Mix the reagents by centrifuging in a microcentrifuge for 1 s.

3. If the thermal cycler does not have a heated lid, add 50 μl of light mineral oil to prevent evaporation during thermal cycling.

4. Place the tube in a thermal cycler and program for the following temperature regime:
   (a) 94°C[3] for 5 min (to denature the template);

(b) 94°C for 1 min[4]; ✍

(c) 55°C[6] for 1 min[4]; repeat 25–35 times[7]

(d) 72°C for 1 min[5]; ✍

(e) 72°C for 2 min (to ensure all molecules are completely synthesized).

5. Samples can be left in the thermal cycler and held at room temperature or 4°C until you are able to remove them for further processing. Generally room temperature is sufficient although some protocols may require a low temperature. It is not a good idea to routinely cool the samples at 4°C for extended periods if this is not necessary as this will reduce the lifetime of the thermal block in the thermal cycler.

6. Remove the tube from the thermal cycler. If the samples are overlaid with mineral oil, carefully insert a pipette tip under the layer of mineral oil and remove about 45 µl of the reaction taking care not to remove any mineral oil.

7. Wipe the outside of the pipette tip with tissue to remove mineral oil sticking to the tip then transfer into a fresh tube.

8. Analyze between 5 and 15 µl of the sample on an agarose gel using suitable DNA molecular size markers as described in Chapter 5.

NOTES

1. PCR buffers are generally supplied by the manufacturer when you purchase a thermostable DNA polymerase. Check the composition of the buffer and specifically whether it contains $MgCl_2$. Magnesium ions are critical for DNA synthesis. Some buffers will contain $MgCl_2$, typically designed to give a final concentration of 1.5 mM in the final PCR. Other buffers will not contain any $MgCl_2$, but a stock solution will usually be supplied by the manufacturer to allow you to determine the optimal $MgCl_2$ concentration.

2. If you are setting up several reactions then prepare a premix of any common components to reduce pipetting steps and potential contamination, as described in Chapter 4. There is a useful online form at http://www.sigmaaldrich.com/Area_of_Interest/Life_Science/ Molecular_Biology/PCR/Key_Resources/PCR_Tools.html, for calculating the amounts of reagents for premixes. Remember to add 1 or 2 additional reactions to account for pipetting inaccuracies.

3. The denaturation temperature should be as low as reasonable to denature the template DNA and often 92°C will be efficient, although most protocols will recommend 94°C, and most people use this temperature. For difficult templates, such

as GC-rich sequences, a higher temperature may be necessary, perhaps 96°C. Also this extended initial denaturation phase may not be necessary or could be significantly reduced to 1 or 2 min in many applications. These measures will extend the functional life of the DNA polymerase molecules.

4. The length of incubation times at each step will depend critically on the thermal cycler characteristics. Often short times of 10–30 s are sufficient for the denaturation and annealing steps. In robust PCR screening for thermal cyclers that monitor tube temperature (Chapter 3) the incubations can be as short as 1 s.

5. The time for the extension step is usually based on the rule of thumb of 1 kb min^{-1}. For shorter products therefore the time can be reduced, while for longer templates it should be increased.

6. This annealing temperature of 55°C is a useful starting point for many PCRs, but can optimally be between 40 and 72°C, depending upon the primer–template combination.

7. The number of cycles depends upon the complexity and amount of template added. Generally for plasmid templates 25 cycles is sufficient whereas for genomic DNA between 30 and 35 cycles are usually necessary. It is sometimes helpful during a genomic amplification to remove 5 μl aliquots at 30 and 35 cycles to compare with the 40-cycle sample to follow the accumulation of the specific band.

Reagents and instrumentation

3

3.1 Technical advances in PCR

The major technical advances that have allowed PCR to become such a routine and accessible tool are:

- thermostable DNA polymerases (*Table 3.2*; Sections 3.10–3.15);
- automation of the temperature cycling process (Section 3.19).

Today PCR is a technically simple operation in which reagents are mixed and incubated in a thermal cycler that automatically regulates the temperature of the reaction cycles according to a preprogrammed set of instructions. The DNA polymerase, being thermostable, need only be added at the start of the reaction so once you have started your PCRs you can get on with another experiment! This Chapter deals with the reagents required for PCR including buffer components, oligonucleotide primer design, thermostable DNA polymerases and template preparation, before dealing with thermal cyclers for performing PCR.

3.2 Reagents

Always remember to thoroughly thaw out and mix buffer and dNTP solutions. If you only partially thaw a solution then differential thawing of components will mean you are not adding the correct concentrations of reactants to your PCR. As a routine approach place the tube in an ice bucket some time before you are going to set up the PCRs. Allow the solution to thaw, vortex briefly, or for a small volume flick the tube with your finger. Place the tube in a microcentrifuge and briefly (1 s) centrifuge to collect the mixed contents at the bottom. Similarly with enzyme solutions, which will not freeze at –20°C due to the glycerol concentration, you should flick them and spin briefly to mix and collect at the bottom of the tube before taking an aliquot to add to your PCRs.

3.3 PCR buffers

Most suppliers of thermostable DNA polymerases provide 10× reaction buffer with the enzyme. Otherwise the following general 10× buffer produces good results with *Taq* DNA polymerase:

- 100 mM Tris-HCl (pH 8.3 at 25°C);
- 500 mM KCl;
- 15 mM $MgCl_2$;
- 1 mg ml^{-1} gelatin;

- 0.1% Tween-20;
- 0.1% NP-40.

The buffer solution should be autoclaved prior to addition of the nonionic detergents (Tween-20 and NP-40), then aliquoted and stored at –20°C. Some buffer recipes recommend including BSA (bovine serum albumin) at 500 µg ml^{-1}.

Tris.HCl

Tris.HCl is a dipolar ionic buffer and the pH of a Tris buffer varies with temperature so during PCR the pH will vary between about 6.8 and 8.3. In fact *Taq* DNA polymerase has a higher fidelity at the lower pH values that occur at the higher temperatures of PCR. It has been recommended that buffers such as Bis–Tris propane and Pipes would be more useful for high fidelity PCR as they have a pKa between pH 6 and 7 and the pH of solutions containing them do not change as significantly with temperature (1).

KCl

KCl can assist primer–template annealing although at high concentrations this can go too far and it may lead to anomalous products through the stabilization of mismatched primers to nontarget sites.

Magnesium

Magnesium is one of the most critical components in the PCR as its concentration can affect the specificity and efficiency of the reaction. *Taq* DNA polymerase is dependent upon the presence of Mg^{2+} and shows its highest activity at around 1.2–1.3 mM free Mg^{2+}. Standard PCR buffers, such as the one shown above, contain 1.5 mM $MgCl_2$; however, buffers for enzymes such as *Pwo* DNA polymerase (Section 3.12) contain 2 mM $MgSO_4$ and not $MgCl_2$. The free Mg^{2+} concentration is affected by the dNTP concentration. There is equimolar binding between dNTPs and Mg^{2+}.

For example, if each dNTP were present at a concentration of 200 µM, the total [dNTP] = 800 µM. The free [Mg^{2+}] = 1 500 – 800 = 700 µM and this is significantly below the optimal concentration for *Taq* DNA polymerase. However, if each dNTP was present at a concentration of 50 µM, the total dNTP concentration = 200 µM. The free Mg^{2+} concentration = 1 500 – 200 = 1 300 µM which represents the optimal concentration for *Taq* DNA polymerase. The magnesium concentration can also affect the fidelity (error rate) of DNA polymerases (Section 3.11). With excess magnesium *Taq* DNA polymerase is more error-prone than at lower concentrations. *Protocol 2.1* should represent a good compromise between yield and fidelity and is a reasonable starting point. If results are not as expected, then perform a Mg^{2+} optimization experiment. Note that with proofreading DNA polymerases the dNTP concentration should not be lower than 200 µM for each dNTP to guard against nuclease activity degrading primers (Sections 3.4 and 3.12).

Suppliers of thermostable polymerases may supply their enzymes with a buffer that lacks magnesium and a magnesium stock solution to allow the user to optimize the magnesium concentration most appropriate for their application. Do not make the common mistake of assuming that magnesium is in every buffer supplied. It is also possible to obtain a variety of buffers and additives to optimize conditions for PCR. For example, Stratagene produce an Opti-Prime™ PCR optimization kit comprising 12 different buffers and 6 additives, allowing a range of buffer conditions to be tested. Once optimized conditions have been determined the appropriate buffer can be purchased separately. Epigene also produce a Failsafe PCR optimization kit comprising a range of buffers.

3.4 Nucleotides

Stock solutions of dNTPs can be purchased from many commercial sources and it is recommended that you use such ready prepared solutions, as these are quality assured. Stock solutions (100–300 mM) should be stored at −70°C and working solutions should be prepared by diluting stocks to between 50 μM and 200 μM of each dNTP in sterile double-distilled water. Because these working solutions should ideally only be stored for 2–3 weeks at −20°C it is recommended that relatively small volumes of working solutions are made. It is important for successful PCR that the four dNTPs are present in equimolar concentrations otherwise the fidelity of PCR can be affected. Similarly, the concentration of dNTPs should be around 50–200 μM. If the concentration is higher the fidelity of the process will be adversely affected by driving *Taq* DNA polymerase to misincorporate at a higher rate than normal, while if the concentration is lower it may affect the efficiency of PCR. Protocols often suggest using 200 μM of each dNTP. This amount would be sufficient to synthesize about 10 μg of product although the most you are likely to achieve is 2–3 μg. Reducing the concentration of dNTPs below 200 μM each is not recommended when proofreading polymerases are being used as they have a 3′→5′ exonuclease activity that will degrade single-stranded DNA molecules such as the primers (Section 3.12). This activity increases as nucleotide concentration decreases. *Taq* and other thermostable DNA polymerases will usually incorporate modified nucleotides into DNA.

3.5 Modified nucleotides

Various modified nucleotides can be incorporated into products during PCR amplifications for various purposes including:

- secondary structure resolution:
    - 7 deaza-dGTP reduces secondary structure in G-rich regions of DNA to improve PCR or sequencing;
- prevention of contamination:
    - dUTP can be used to replace dTTP to provide a substrate for uracil *N*-glycosylase to allow destruction of previously amplified PCR products to prevent carryover (Chapter 4);

- radiolabeling of PCR products:
  - $[\alpha^{32}P]dNTPs$;
  - $[\alpha^{33}P]dNTPs$;
  - $[\alpha^{35}S]dNTPs$;
- nonradioactive labeling of PCR products:
  - usually the labels are modified forms of dUTP carrying biotin, fluorescein or digoxigenin and are substituted for some of the dTTP in the reaction mix (for example 50 μM modified dUTP + 150 μM dTTP). Bromodeoxyuridine can also be used;
- DNA sequencing:
  - ddNTPs as chain terminators in standard sequencing;
  - fluorescently labeled ddNTPs in fluorescent DNA sequencing (Chapter 5);
- random mutagenesis:
  - modified nucleotides eg. dPTP and 8-oxo-dGTP (Chapter 7).

3.6   PCR premixes

Increasingly PCR premixes are becoming available. These contain buffer, dNTPs and *Taq* DNA polymerase as a premixed reagent at a concentration that allows addition of template DNA and primers to produce the final reaction volume. In some cases the buffers contain no magnesium, allowing optimization experiments to be undertaken by addition of magnesium stocks. It is also possible to obtain custom prepared stocks with desired concentrations of reagents, such as magnesium, optimized for your experimental procedure. Clearly the use of premixes is highly advantageous for high-throughput screening or template preparation applications, particularly with the increasing use of automated robotics for reaction set-ups in, for example, clinical screening and genomics laboratories. However, it is also worth considering the use of premixes for more routine applications. Many manufacturers now provide premix reagents for both standard and real-time PCR applications available as bulk reagents or prealiquoted into PCR plate format.

3.7   Oligonucleotide primers

Oligonucleotides are widely available and there are many companies (such as Alpha DNA, Biosource, Bio-Synthesis, Integrated DNA Technologies, Invitrogen, Midland Certified Reagent Company, MWG Biotech, PE Biosystems and Sigma Genosys) that offer low-cost custom synthesis and purification of your primer sequences within a few days of ordering. For most PCRs (with the exception of some genomic mapping approaches, such as RAPD analysis, Chapter 11) you will need two primers of different sequence that anneal to complementary strands of the template DNA. When you know the DNA sequence of your template it is quite easy to design suitable primers to amplify any segment that you require. There are several computer programs that can be used to assist primer design. Web primer (http://seq.yeastgenome.org/cgi-bin/web-primer); Primer3 (http://frodo.wi.mit.edu/cgi-bin/primer3/primer3_www.cgi); Oligoperfect designer (http://www.invitrogen.com/content.cfm?pageid=9716); Fastpcr (http://www.biocenter.helsinki.fi/bi/Programs/fastpcr.htm); Net primer (http://premierbiosoft.com/netprimer/index.html. However, in practice many people still design primers by following some simple rules.

A primer should:

- be 16–30 nucleotides long, which provides good specificity for a unique target sequence, even with a starting template as complex as human genomic DNA;
- contain approximately equal numbers of each nucleotide;
- avoid repetitive sequences or regions containing stretches of the same nucleotide as this can lead to 'slipping' of the primer on the template;
- avoid runs of three or more G or Cs at the 3'-end as this can lead to mispriming at GC-rich regions;
- not be able to form secondary structures due to internal complementarity;
- not contain sequences at the 3'-ends that will allow base pairing with itself or any other primer that it may be coupled with in a PCR; otherwise this can lead to the formation of primer-dimers.

A primer-dimer is the product of primer extension either on itself or on the other primer in the PCR as shown in *Figure 3.1*. Since the primer-dimer product contains one or both primer sequences and their complementary sequences they provide an excellent template for further amplifications. To make matters worse smaller products are copied more efficiently (and a primer-dimer is about as small as you can get!); primer-dimers can dominate the PCR and sequester primer from the real target on the template DNA.

In many cases the primer sequence does not need to be a perfect complement to the template sequence. The region of the primer that should be perfectly matched to the template is the 3'-end because this is the end of the primer that is extended by the DNA polymerase and is therefore most important for ensuring the specificity of annealing to the correct target sequence (*Figure 3.2*). In general at least the first three nucleotides at the 3'-end should perfectly match the template with complemarity extending to about 20 bp with a few mismatched bases. The 5'-end of the primer is less important in determining specificity of annealing to the target sequence and this means it is possible to alter the sequence in some desirable manner to facilitate subsequent cloning, manipulation, mutagenesis, recombination or expression of the PCR product (*Figure 3.2*). A common modification is to introduce a restriction site so that the amplified product can be cloned into the desired plasmid vector simply and efficiently. A restriction endonuclease site can simply be added close to the 5'-end of the primer (Chapter 6) or it can be generated within the primer region by altering one or more nucleotides (Chapter 7).

Longer additions can be made to the 5'-end of a primer including promoter sequences to allow *in vitro* transcription of the PCR product, or sequences to allow the splicing or joining of PCR products (Chapter 7). A range of mutations can be introduced into a PCR product by altering the sequence of the primer (Chapter 7). The primers define the region of DNA to be amplified and can be used to tailor the PCR product for subsequent use.

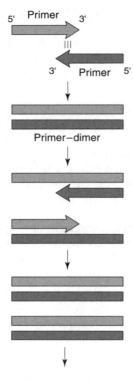

Figure 3.1

Primer-dimer formation is due to self-priming by one or both primers and can be overcome by careful design of primers to try to ensure they do not have complementary 3'-ends. If one or both of the primers in the PCR anneal because their 3'-ends have some complementarity, then during PCR the primers self-prime resulting in a primer-dimer. At the next PCR cycle, each primer-dimer strand can act as a new template resulting in highly efficient amplification of this small artifact product.

Figure 3.2

The 3'-region of a primer is critical for efficient annealing to the correct target sequence. The 5'-region is less important and can be modified to carry additional sequences, such as restriction sites or promoter sequences, that are not complementary to the template.

Melting temperature (T_m)

You will often see references to the melting temperature of a primer as an indicator of the annealing temperature step during PCR. The T_m is the temperature at which half the primers are annealed to the target region. There are a number of approaches for calculating T_m. The simplest method for primers up to about 20 nucleotides in length is based on adding up the number of each nucleotide in the primer then using the formula [1]:

$$T_m = ((\text{Number of G+C}) \times 4°C + (\text{Number of A+T}) \times 2°C) \qquad [1]$$

This formula reflects the fact that G/C base pairs are more stable than A/T base pairs due to their greater hydrogen bonding. It can provide a rough guide to choosing primer sequences that have similar T_ms. Originally it was devised for hybridization assays in 1 M salt, an ionic strength significantly higher than that used in PCR (2). It is best when designing a pair of primers to try to match their T_ms so that they will have similar annealing temperatures. It is obviously not very appropriate to use one primer with a T_m of 40°C and another with a T_m of 68°C, for example. If you use the formula above to calculate T_m then it is probably best to set the annealing temperature in the first PCR to about 5°C below the calculated T_m.

A more accurate formula [2] that can be used for oligonucleotides between about 15 and 70 nucleotides in length in aqueous solution (3) is:

$$T_m = 81.5 + 16.6(\log_{10}(I)) + 0.41(\%G+C) - (600/N) \qquad [2]$$

where I is the concentration of monovalent cations and N is the length of the oligonucleotide.

An alternative approach [3] for primers 20–35 nt long is to calculate T_p, the optimized annealing temperature ± 2–5°C (4):

$$T_p = 22 + 1.46 ((2\times \text{ number of G+C}) + (\text{number A+T})) \qquad [3]$$

The only region that you need to consider when calculating a T_m (or T_p) is that part of the primer that will anneal to the template; if you have added a long tail at the 5′-end then you can forget about this.

As an example let us look at the following primer sequence annealed to its complementary template:

```
5′-AGTTGCT GAATTC GTGAGTCCCTGAATGTAGTG-3′
        |  |  |          ||||||||||||||||||||
3′-TAGCTCGCTAGGGTCGGTCCACTCAGGGACTTACATCACGATCGTTTGCAATCCCATA-5′
```

The primer is designed to contain a tail including the site for the restriction enzyme *Eco*RI (GAATTC), shown boxed. This tail does not contribute to the specificity of the primer annealing to its target sequence and so we only need consider the 20 nucleotides at the 3′-end of the primer when determining the T_m. This region of the primer contains 7Gs, 3Cs, 4As and 6Ts.

According to formula (1), the $T_m = (10 \text{ (G+C)} \times 4°C) + (10 \text{ (A+T)} \times 2°C) = 60°C$. According to formula [2], assuming a standard monovalent ion concentration of 50 mM (KCl), the $T_m = 81.5 + 16.6(\log_{10}(0.05 \text{ M})) + 0.41(50) - (600/20) = 50.4°C$. According to formula [3], the $T_p = 22 + 1.46 (20 + 10) = 65.8 ± 2–5°C$.

As you can see there is significant variation in calculated values. Such calculations only provide guidelines for the annealing temperature to use in PCR. In practice it is usually necessary to determine the optimum annealing temperature empirically. Ideally you should use the highest annealing temperature that gives you efficient amplification of the desired product with the lowest level of nonspecific product. In some cases it is possible to perform two-step PCRs where the annealing temperature of 72°C is also the temperature for optimum DNA synthesis. The optimization of annealing temperatures is greatly simplified if you have access to a thermal cycler with gradient heat block facility (Section 3.19).

5'-end labeling of primers

PCR products can be cloned directly into various vectors, but unless you are performing ligation independent cloning, the primer or PCR product must be 5'-phosphorylated to allow formation of a phosphodiester bond during ligase-mediated joining with the vector. When they are chemically synthesized primers will not contain a 5'-phosphate group unless this has been requested. Phosphoramidites are available for addition of 5'-phosphate groups during oligonucleotide synthesis but this can be expensive. In the lab the process of phosphorylating the primer, or indeed the PCR product, is relatively simple and involves treatment with T4 polynucleotide kinase and ATP (*Protocol 3.1*). The γ-phosphate group of ATP is transferred to the 5'-OH of the unphosphorylated primer. The same process is used to end-label a primer with ^{32}P by transfer from [γ-^{32}P]ATP allowing autoradiographic detection of the PCR product in experiments such as DNA shift assays, protein binding site determinations or direct analysis of PCR products. Such a 5'-end label can also be useful for determining whether a restriction enzyme has successfully cleaved a PCR product (Chapter 6), as the label will be lost from the product upon cleavage.

It is also possible to introduce a number of other labels that facilitate PCR product detection, localization, quantification and isolation. A widely used method for labeling primers that is useful not only for detection but also for purification, is biotinylation. There are now several biotin phosphoramidite reagents that allow simple and convenient 5'-end labeling and these are readily available from commercial oligonucleotide custom synthesis suppliers and other companies. Biotin can be detected by using streptavidin, which is widely available in a number of forms including enzyme-linked systems for nonisotopic detection, and even associated with paramagnetic particles for simple capture and purification of PCR products (Chapters 5 and 6).

Another nonisotopic labeling method widely used for nucleic acid detection is digoxigenin, which can be coupled to primers that are synthesized with a 5'-AminoLink (*Figure 3.3*). In addition to their incorporation as end-labels in PCR primers, both biotin and digoxigenin can also be incorporated into PCR products as nucleotide analogues during the PCR as described later (Chapter 5).

Fluorescent dye-labeled primers can be produced for use in laser detection of product accumulation in real time (Chapter 9), in some DNA sequencing

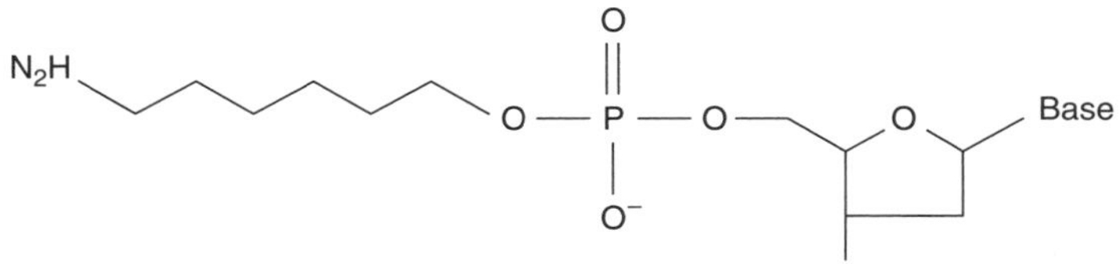

Figure 3.3

Structure of AminoLink attached to the 5′-deoxyribose of an oligonucleotide. The reactive amine group is separated from the DNA by a spacer.

approaches (Chapter 5) and for analysis of genomic polymorphisms (Chapter 11). Again many fluorescent dyes are available in an active ester form, for example N-hydroxysuccinimide (NHS), and can be coupled to AminoLink-oligonucleotides. There are also a variety of fluorescent dye phosphoramidites such as FAM (6-carboxyfluorescein), HEX (4,7,2′,4′,5′,7′-hexachloro-6-carboxyfluorescein), ROX (6-carboxy-X-rhodamine) or TET (tetrachloro-6-carboxyfluorescein) that can be incorporated at the 5′-end of the primer during chemical synthesis by a number of oligonucleotide supply companies.

Non-nucleosidic phosphoramidites are also now available and can be incorporated into PCR primers. These compounds, such as naphthosine R (www.DNA-technology.dk) (usually two contiguous naphthosines are needed), are not recognized as normal nucleotides but act to terminate the DNA polymerase. This can result in the production of double-stranded PCR products with single-stranded tails that can be subsequently used for detection or isolation purposes (*Figure 3.4*).

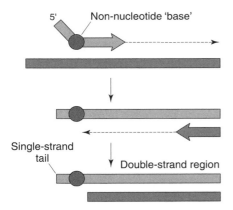

Figure 3.4

Incorporation of a non-nucleosidic phosphoramidite within a primer allows the production of a PCR product with a single-strand tail because the DNA polymerase terminates at the non-nucleosidic 'base'.

Degenerate primers (mixtures of primers)

Primers for PCR are usually a unique sequence designed from the known DNA sequence of the template. However, for certain applications you may not know the sequence of the template DNA. This situation normally arises when the gene sequence is not known, but amino acid sequence data are available from the protein encoded by the target gene (Chapter 10). In such cases there are two options. If many genes have been sequenced from the genome of the organism in question then it is possible to generate a codon usage table or access http://www.kazusa.or.jp/codon/ and to identify the codons that the organism uses most frequently for each amino acid. This would allow you to generate a 'best guess' at the likely DNA sequence that would encode the known peptide sequence, so that you could synthesize a single oligonucleotide sequence as a primer. Of course this assumes that your guess is reasonably correct. If the gene happens to use different codons from those most frequently used by the organism then you risk never amplifying the target gene. The second approach is to use a mixture of different oligonucleotides where all the possible codons for each amino acid are present. The degeneracy of the genetic code means that a single amino acid may be encoded by several possible codons. Thus a given peptide sequence might be encoded by several possible DNA sequences and it is necessary to synthesize a mixture of all the possible DNA sequences of the primer that correspond to the region of peptide sequence. It may however be possible to combine the two approaches to reduce the complexity of a degenerate primer mixture by identifying very pronounced codon bias and including such codons as unique rather than degenerate sequences. Such primers are called degenerate primers and there is further discussion of their use in Chapter 10.

Figure 3.5 illustrates the design of degenerate primers from an amino acid sequence. Two examples are shown that differ in the way positions that could be any of the four dNTPs are handled. In the first example (Primer 1), a mixed base synthesis is performed with all four dNTPs added to the growing oligonucleotide resulting theoretically in 25% of the molecules having an A, 25% G, 25% C and 25% T. In the second example (Primer 2), such positions are substituted by one nucleotide, deoxyinosine (I), which is capable of pairing with all four bases. This reduces the complexity of the oligonucleotide mixture. Deoxyinosine is a widely used universal base although its capacity to pair with the four bases is not equal. Universal bases (5,6) are also available as phosphoramidites for use in primer synthesis that base pair equally with all four bases. Although universal bases are useful, care should be taken when using multiple deoxyinosines in that the higher the degeneracy the more mismatches, ultimately resulting in higher background and nonspecific amplification. Ideally the three nucleotides at the 3'-end of the primers should be perfectly matched with the template.

The two main objectives when designing a degenerate primer are to have the primer as long as possible and to have the lowest possible degeneracy (the number of nucleotides needed to cover all combinations of nucleotides). This can at times be problematic but by following some simple rules the task is made easier. First, identify an eight to ten amino acid stretch in your protein that is rich in amino acids encoded by only one or two codons (Met, Trp, Phe, Cys, His, Lys, Asp, Glu, Gln, Asn, Tyr) and that has no or few amino acids encoded by six codons (Ser, Leu, Arg). Once the amino acid sequence has been defined translate it to nucleotides based on the respective codons. It can sometimes be useful to translate the sequence using the IUPAC symbols for degeneracy (*Table 3.1*) as you will need these when you order your oligonucleotide from a commercial company. Once you have identified a suitable region the primer sequence can be refined to reduce the degeneracy. Degeneracy is calculated by multiplying all the degeneracy values in the primer together. Remember that synthetic oligonucleotides are chemically synthesized in a 3'→5' direction so it is important to avoid a degenerate position at the 3' terminal end of the primer. If we take Primer 1 in *Figure 3.5* as an example, the removal of the terminal degenerate nucleotide position from the glycine codon at the 3'-end of the primer provides a valuable 3'-GG clamp (*Figure 3.5*). Another way of reducing degeneracy is to make the third nucleotide at the 5'-end of the primer fixed, as tight annealing is less important at the 5'-end. An informed decision can be made using codon usage information for the organism in question. If we apply this rule to Primer 1 in *Figure 3.5* the degeneracy decreases from 256 to 64, which is a huge improvement.

Although T_m calculations have been described earlier in this Section, determining the T_m for degenerate primers is slightly different, bearing in mind that as the primer concentration decreases the T_m also decreases. Over a wide range of primer concentrations the T_m decreases by 1°C for each two-fold decrease in primer concentration. This implies that if the degeneracy of a primer is 1000 the melting temperature should be approximately 10°C lower than that calculated without the primer concentration corrected. If the specific primer concentration is low due to high degeneracy this can simply be overcome by increasing the overall primer concentration.

Although degenerate primers can be designed manually following the above rules there are several web-based programs (COnsensus-DEgenerate Hybrid Oligonucleotide Primers: http://blocks.fhcrc.org/codehop.html and GeneFisher: http://bibiserv.techfak.uni-bielefeld.de/genefisher/) that will design your degenerate primers after you input your amino acid sequence.

Primer concentration

The amount of primer that is used in a PCR depends upon the experiment, as the primer to template ratio is an important consideration. Generally, the two primers should be used at equal concentrations and recommended

Table 3.1 When ordering a degenerate oligonucleotide primer from a commercial company the following IUPAC symbols for degeneracy are used

IUPAC symbols	R	Y	M	K	S	W	H	B	V	D	N
Nucleotides	A	C	A	G	C	A	A	C	A	A	A
	G	T	C	T	G	T	C	G	C	G	C
							T	T	G	T	G
											T

Amino acid sequence   **A   D   T   E   W   D   G   G**
Possible DNA sequences   NNNGCAGACACAGAATGGGACAAAGGANNNN
                            G   T   G   G   T   G   G
                            C   C   C
                            T   T   T

Primer 1 Mixed base synthesis (256 different sequences)

           GC**A**GACAC**A**GAATGGGACAAAGG
     5′   **G   T   G   G   T   G**   3′
                C   **C**
                T   **T**

Primer 2 Universal base synthesis (8 different sequences)

           GC**I**GACAC**I**GAATGGGACAAAGG

Figure 3.5

Example of the design of degenerate primers by back-translation from amino acid sequence data. The mixed base synthesis version includes all four nucleotides at positions of four-fold degeneracy in the Ala and Thr codons and leads to a mixture of 256 different oligonucleotide sequences in the primer sample. The universal base option replaces these four-fold degenerate positions with the single base deoxyinosine or other universal base thereby making the primer sample less complex with only eight different sequences. Both primer samples should be capable of priming on the DNA sequence that encoded the amino acid sequence shown. Note the 3′-end of the primers corresponds to positions 1 and 2 of the Gly codon which provides two unique 3′-nucleotides.

amounts vary from 0.1 μM to 1 μM which are equivalent to 5–50 pmol of each primer in a 50 μl reaction volume. It has been calculated that at the lower concentration of 0.1 μM for a genomic DNA amplification the primer excess over template is around 10^7 (*Table 2.2*) and remains fairly constant throughout the PCR because 95% of the primers remain unused after a 30-cycle reaction. It is suggested that using high concentrations of primers can lead to artifacts with increasing likelihood of primer-dimer formation and mispriming on nontarget sequences.

When your primer arrives it should contain information about the quantity that is provided and so it should be straightforward to dissolve the sample in an appropriate volume of water. Of course this may not always be the case and there may be times when you need to determine the concentration of an oligonucleotide sample.

To calculate the concentration of an oligonucleotide the formula $A_{260} = \varepsilon c l$ can be used.

- A_{260} is the absorbance at 260 nm of an aliquot of the primer; if necessary dilute the primer so that the absorbance value is within the range 0.1–0.8, then multiply the A_{260} value by the dilution factor. For example if you dilute 10 μl of primer into the 1 ml sample, multiply the A_{260} value you measure by 100, because you diluted the original sample 100-fold to make the measurement.
- ε is the molar extinction coefficient ($M^{-1}cm^{-1}$). You can calculate this quite precisely or fairly crudely.

If you want to be precise then you add up the number of each of the four nucleotides in the primer and multiply each number by the appropriate ε value (15 200 for A, 8 400 for T, 12 010 for G and 7 050 for C). For example the 20 mer GTGAGTCCCTGAATGTAGTG would have an ε = (15 200 × 4) + (8 400 × 6) + (12 101 × 7) + (7 050 × 3) = 216 420. The crude approach uses an average ε value of 10 600 for each nucleotide in the sequence, which assumes that each of the four nucleotides are present in equal numbers. For the oligonucleotide above this would give a value of 212 000, which is in good agreement with the more precise value. Of course if the oligonucleotide contained predominantly G and A or C and T then the agreement would not be as good.

- c is the concentration of the primer (M) that we are trying to calculate.
- l is the pathlength (cm) which is usually 1 (one).

So to calculate the primer concentration from $c = A_{260}/\varepsilon l$ we need to measure A_{260}, calculate ε and know l.

Taking the example of the primer above, if we dilute 10 μl of stock solution into 1 ml water in a 1 cm pathlength cell and measure an absorbance of 0.15 we can calculate the concentration of oligonucleotide in the original sample:

$$c = (0.15 \times 100 \text{ [the dilution factor]})/(216\ 420 \times 1) = 0.000069 \text{ M}$$
$$= 69 \text{ μM}$$
$$= 69 \text{ pmol μl}^{-1}$$

Another way to determine primer concentration is to determine the A_{260} of a diluted sample ($A_{260} = 1 = 33$ μg ml^{-1} single-stranded DNA) and combine with the length of the oligonucleotide N (in this case 20 nt) and molecular mass of an average nucleotide (dNMP; 325 Da). If we take the solution above that gives an $A_{260} = 0.15$ for a 10 μl sample of the 20-mer diluted into 1 ml then this is equivalent to an A_{260} of 15 (0.15 × 100; the dilution factor) for our stock solution. So by using the formula $c = A_{260}$ (33 μg/N × 325) we can calculate the concentration c of the stock solution:

$$c = (0.15 \times 100) \times (33 \text{ μg}/ (20 \times 325)) = 15 \times 0.005 = 0.076 \text{ μmol ml}^{-1}$$
$$= 76 \text{ nmol ml}^{-1}$$
$$= 76 \text{ pmol μl}^{-1}$$

This value is in reasonable agreement with the 69 pmol μl^{-1} value determined from the extinction co-efficients. The solution can be diluted to give a working stock of perhaps 10 μM (10 pmol μl^{-1}). If you are going to repeatedly use a particular primer pair together, then it can be useful to mix these as a working stock. Be careful when handling stock primer solutions that you do not contaminate the stock with another primer or any extraneous DNA (Chapter 4).

Other useful conversion factors for primers include:

- calculation of pmol to ng = (no. of pmol × N × 325)/1 000
  for example, 2.5 pmol of a 20-mer = (2.5 × 20 × 325)/ 1 000
    = 16.3 ng
- calculation of ng to pmol = (ng × 1 000)/(N × 325)
  for example 70 ng of the 20-mer = (70 × 1 000)/(N × 325)
    = 10.8 pmol

Primer stocks

Oligonucleotides will usually arrive either in a dried-down state, in aqueous solution, or in ammonia, depending upon your supplier. Commercial suppliers are most likely to supply dried or aqueous solution stocks while in-house facilities may provide ammonia solutions. Redissolve the oligo-nucleotides to an appropriate concentration, perhaps 10–100 pmol μl^{-1} in water or 10 mM Tris-HCl, pH 8.0, and take a sample for use as a working stock while storing the remainder at –20 or –70°C, or by re-drying and storing in a dried state. If the stock arrives in aqueous solution then deal with it similarly. If it arrives in ammonia, then store the stocks at –20°C, remove an aliquot, dry it down in a centrifugal vacuum dryer then redissolve in water or 10 mM Tris-HCl (pH 7.5–8) as for example a 10–100 pmol μl^{-1} working stock solution. Working stock solutions may be stored for many weeks at –20°C.

3.8 DNA polymerases for PCR

During DNA synthesis the DNA polymerase selects the correct nucleotide to add to the primer to extend the DNA chain according to the standard Watson and Crick base pairing rules (A:T and G:C).

Two classes of DNA polymerase are commonly used according to the template they copy:

- DNA-dependent DNA polymerases;
- RNA-dependent DNA polymerases also called reverse transcriptases.

A DNA polymerase always catalyses the synthesis of DNA in the 5'→3' direction. Some DNA polymerases also have a 3'→5' exonuclease activity, called a 'proofreading' activity which 'checks' that the correct base has been added to the growing DNA strand. When an incorrect nucleotide is incorporated the proofreading activity will remove the incorrect base so that the synthesis activity can incorporate the correct base. This correction mechanism increases the accuracy, or fidelity, with which the polymerase copies the template strand. We will consider the fidelity of the polymerase reaction in Section 3.11 when we consider *Taq* DNA polymerase.

When different DNA polymerases are being compared, there are two overall properties that are important for PCR; the fidelity and the efficiency of synthesis. The efficiency is a consequence of processivity and rate of synthesis; although it can be difficult sometimes to distinguish between these aspects. Processivity is a measure of the affinity of the enzyme for the template strand. The stronger the interaction, the more processive the polymerase should be, and so the more DNA it will synthesize before it dissociates from the template. A DNA polymerase molecule dissociates from the template quite often, perhaps after copying as few as 10 bases, or as many as 50 or 100. After a DNA polymerase molecule dissociates, the 3'-end of the newly synthesized DNA strand is a substrate for another DNA polymerase molecule which can associate with the template and synthesize another stretch of DNA. This process continues until the complete DNA strand has been synthesized. The *rate of synthesis* or speed at which a polymerase copies the template, usually measured as nucleotides incorpo-rated per second, also varies. Some enzymes incorporate only 5–10 nt s^{-1}

while others incorporate more than 100 nt s^{-1}. The properties of some DNA polymerases commonly used in PCR are compared in *Table 3.2.*

3.9 Early PCR experiments

The first PCR experiments used the Klenow fragment of DNA polymerase I (Klenow fragment) from *E. coli,* an enzyme still widely used in molecular biology experiments. The Klenow fragment has two enzyme activities, 5'→3' DNA synthesis activity and 3'→5' exonuclease (proofreading) activity. Although PCR using the Klenow fragment was successful in amplifying DNA it had severe drawbacks:

- the operator had to be present throughout the reaction to add regularly fresh Klenow fragment as the high temperature used for the denaturation step also denatured the enzyme, so an aliquot had to be added at each 37°C DNA synthesis step;
- due to this continual addition, large amounts of Klenow fragment were required, making the process expensive;
- the low temperature of 37°C needed for the DNA synthesis steps led to efficient amplification of nontarget regions of DNA because primers could anneal to nontarget regions.

You can probably imagine how boring and tedious this early process must have been. Only a few reactions could be performed at any one time and as you probably know, if you carry out any sort of monotonous task it is quite easy for your mind to wander and to make a mistake – such as incubating the tubes at the wrong temperature, or forgetting to add enzyme!

3.10 Thermostable DNA polymerases

Taq DNA polymerase, from the thermophilic bacterium *Thermus **aquaticus**,* was first described by Brock and Freeze in 1969 (7) but was not widely used until the need for such an enzyme arose for PCR. The stability of *Taq* DNA polymerase at the high temperatures used in PCR allowed repeated amplification cycles following the single addition of enzyme at the start of the reaction. The enzyme displays an optimum temperature for DNA synthesis of around 72–75°C. As high temperatures (55–72°C) can be used during the primer annealing steps the added bonus is improved specificity of primer annealing leading to greater amplification of target sequences and less amplification of nontarget sequences.

3.11 Properties of *Taq* DNA polymerase

Taq DNA polymerase is a 94 kDa protein that has two catalytic activities:

- 5'→3' DNA polymerase with a processivity of 50–60 nucleotides and an extension rate of around 50–60 nt s^{-1}, corresponding to around 3 kb min^{-1} at 72°C; and
- 5'→3' exonuclease.

Table 3.2 Properties of some thermostable DNA polymerases

Enzyme	Supplier[a]	Half-life 95°C	Half-life 97.5°C	Half-life 100°C	Processivity	Extension rate nt s^{-1}	5'→3' exonuclease activity	3'→5' exonuclease proofreading	Error rate per bp	Reverse transcriptase activity	DNA termini of products	Magnesium ion optimum
Taq	PE-AB	40 min	10 min		50–60	50	Yes	No	8×10^{-6}	Weak	3'-A	1.5–4 mM
Stoffel	PE-AB	80 min	20 min		5–10	>50	No	No		Weak	3'-A	2–10 mM
Tth	Various	20 min	2 min		30–40	>33	Yes	No		Yes	3'-A[b]	1.5–2.5 mM
Vent™	NEB	400 min		1.8 h	7	>17	No	Yes (No for exo–)	2.8×10^{-6}		>95% blunt	
DeepVent™	NEB	1380 min		8 h	>20	23	No	Yes (No for exo–)	2.7×10^{-6}		>95% blunt	
Pfu	STRA	>120 min			>20	25	No	Yes (No for exo–)	1.3×10^{-6}			
Pwo	Roche			>2 h			No					2 mM MgSO$_4$
UlTma™	PE-AB		50 min				No	Yes	5.5×10^{-5}		Blunt	2 mM MgSO$_4$ + up to 2 mM MgCl$_2$
ACCUZYME	Bioline							Yes	10^{-6}		Blunt	
BIO-X-ACT	Bioline							Yes	5×10^{-6}		3'-A	
KOD HiFi	NOVA				>300	120	No	Yes	10^{-6}		Blunt	2.5 mM MgCl$_2$

[a] PE-AB, Perkin Elmer-Applied Biosystems; NEB, New England Biolabs; STRA, Stratagene; Roche, Roche Molecular Systems; Bioline; NOVA, Novagen.
[b] *rTth* DNA Polymerase XL from PE Biosystems is reported not to add a 3'-A.

The enzyme has a half-life of around 40 min at 95°C that equates to some 50 cycles under standard PCR conditions. It does not have a 3′→5′ exonuclease 'proofreading' activity which means that errors made by the polymerase activity can not be corrected, an aspect we will return to when we consider fidelity of polymerases later in this section. *Taq* DNA polymerase is a good general purpose enzyme for many routine PCR applications and is capable of amplifying products up to 2–4 kbp quite efficiently. The use of DNA polymerase mixtures can reduce the apparent error rate of *Taq* DNA polymerase and can result in longer products being generated (Section 3.15).

Commercial sources of *Taq* DNA polymerases

Many companies sell *Taq* DNA polymerase under a variety of commercial names under license from Hoffman LaRoche who hold the patent rights to the use of the enzyme for PCR. These include Clontech, Eppendorf, Fermentas, Invitrogen, New England Biolabs, Promega, Qbigene, Qiagen, Sigma Aldrich and Stratagene. Technically any enzyme that is not sold under license should not be used for PCR amplification of DNA. However, legal processes take some considerable time to reach a final conclusion. Promega Corporation have challenged the original patents on *Taq* DNA polymerase which may therefore eventually be revoked. Several companies have developed new sources of thermostable DNA polymerases that are also suitable for use in PCR. Patent protection for use of the enzyme expires shortly.

AmpliTaq®

AmpliTaq® is produced by PE Biosystems and is a recombinant version of *Taq* DNA polymerase that is produced in *E. coli*. Recombinant *Taq* DNA polymerase production in *E. coli* facilitates simple recovery of the thermostable enzyme by incorporation of a heating step during purification. The properties of the enzyme are the same as those given above for *Taq* DNA polymerase but as it is from a recombinant source the purity and batch-to-batch reproducibility are higher than for native enzyme. Due to the bacterial source of the recombinant enzyme it is possible that DNA may be present in the enzyme preparation and could lead to contamination of the PCR. It is recommended that either a native *Taq* DNA polymerase or a source of AmpliTaq® certified to have low DNA content (AmpliTaq® LD, Perkin-Elmer Applied Biosystems) should be used for PCR with bacterial DNA templates.

AmpliTaq Gold®

This is a modified form of AmpliTaq® that is supplied in an inactive form that requires a 9–12 min pre-PCR heat step at 94 or 95°C to activate the polymerase. This enzyme provides a convenient approach to hot-start PCR that is used to improve the specificity of PCRs and which is covered in more detail in Chapter 4. Alternatively it can be activated slowly during PCR to combine a hot-start approach with time-released PCR. The benefits include prevention of mispriming and increased yields of the specific product.

Stoffel fragment

The Stoffel fragment of AmpliTaq® DNA polymerase is named after S. Stoffel who was responsible for generating this modified protein. The Stoffel fragment is a 61 kDa protein that lacks the N-terminal 289 amino acid residues of AmpliTaq® and is also produced as a recombinant protein in *E. coli*. The Stoffel fragment lacks the 5'→3' exonuclease activity of AmpliTaq® and is around two-fold more thermostable, which allows PCR to be performed at higher temperatures leading to more specific primer annealing which may be particularly useful for template DNAs with a high G+C content. Compared with the parent AmpliTaq®, Stoffel fragment is less processive (5–10 nucleotides) but the overall DNA synthesis rate is comparable (3 kb min⁻¹ at 70°C) and it is very efficient at producing relatively short PCR products of less than 500 base pairs. Stoffel fragment works over a broader range of magnesium ion concentrations (2–10 mM) which means you do not need to spend so much time optimizing reaction conditions (discussed in Chapter 4). This can also be useful for multiplex reactions (Chapter 11) where more than one set of primers are being used simultaneously. The Stoffel fragment also works best in a lower ionic strength buffer (10 mM KCl) than AmpliTaq (50 mM KCl).

Fidelity of *Taq* DNA polymerase

For most purposes Amplitaq® or an equivalent recombinant thermostable DNA polymerase represents the most appropriate enzyme for PCR. One deficiency however, is the lack of 3'→5' exonuclease 'proofreading' activity. In fact this deficiency has not proved to be as much of a problem as was originally thought and the frequency of base misincorporation, also referred to as accuracy of the DNA polymerase, has been extensively investigated and is reported to be of the order of 1 error in every 10^4 nucleotides incorporated for base pair changes and 1 in every 4×10^4 for frameshifts. Although other studies report lower error rates. The error rate is the reciprocal of accuracy and provides a measure of the mutation rate per base pair duplicated. It is possible to calculate that for a 400 bp target fragment amplified 10^6-fold an error rate of 1 in 10^4 results theoretically in 33% of the product fragments carrying a mutation. This mutational capacity of *Taq* DNA polymerase should be considered during experimental design to ensure that, should they occur, mutations can be identified. For example, if a product is cloned for further studies, then several independent clones may need to be sequenced to identify one without an error. This is reasonably straightforward when you know the DNA sequence of the wild-type target region. However, if you are amplifying a new sequence then you should sequence several clones to derive a consensus sequence for the target region and allow selection of clones that represent this consensus. Ideally clones should be sequenced from independent PCRs to ensure the highest level of certainty that a clone represents the wild-type sequence. The rationale is that identical errors are extremely unlikely to occur in independent PCRs and therefore polymerase-introduced mutations should be more readily identifiable.

There have been several thorough studies of the error rates of a variety of DNA polymerases used for copying DNA templates. Essentially three methods have been used to examine this feature: the reversion assay, the forward mutation assay and gradient gel assays. These methods are obviously appropriate for determination of the error rate of any DNA polymerase whether mesophilic or thermophilic, and whether or not the enzyme has a proofreading activity.

Reversion assay

This system has been widely used for assessing polymerase-induced error rates. It uses an M13mp2 template carrying defined mutation(s) which lead to a known mutant phenotype, in this case colorless plaques. Following DNA synthesis of a single-strand region including the mutated site, by the DNA polymerase being examined, the error rate can be assessed by the number of reversion events leading to a wild-type phenotype, in this case blue coloration on indicator plates. The approach has been used to measure both single-base substitution (8) and single-base frameshift error rates (9). It has been argued that since this method relies upon measuring reversion of specific mutations at only a few sites, it may bias some estimates of error rate since there can be regions of DNA more prone to or more resistant to polymerase-induced errors.

Forward mutation assay

This system provides a mechanism for assessing the range and frequency of errors introduced by a DNA polymerase within a DNA sequence (10). In essence DNA polymerase errors are measured for the synthesis of a single-stranded gap (390 nt) in an otherwise duplex M13mp2 DNA molecule. The gap is opposite the *lacZ* region and loss of function is scored as phenotypic occurrence of colorless or light blue plaques (mutant) compared with dark blue plaques (wild type). Clearly it is also possible to sequence products from both forward and reverse assays which can provide a definitive indication of the number of polymerase-induced errors, including those that do not affect phenotype.

Gradient gel analysis

This technique relies upon a polyacrylamide gel assay system that is able to discriminate between molecules that carry single-base changes, on the basis of their differential mobility through a gel that includes some form of gradient (11). The target sequence is amplified by PCR and one primer has a 5'-tail comprising some 40 G and C nucleotides. This results in a duplex product with one end comprising a GC duplex that is known as a GC-clamp. This region represents a DNA domain highly resistant to denaturation compared with the target region to which it is attached. As the target sequence migrates through the gel and moves into higher denaturant or temperature it begins to denature and eventually completely denatures resulting in a 'Y-shaped' DNA where the stalk of the Y is the GC-clamp and the tails are the single strands of the target region. This Y-shaped molecule becomes entangled in the gel network and essentially stops migrating. Since single-base changes can dramatically alter the temperature, or denaturant concentration, at which the target sequence will 'melt', there is normally a significant difference in melting characteristics between a wild type and mutant that differ by only one base pair, leading to good separation on the gel. In the first examples a denaturing gradient (DGGE) was used but there is now a system that uses a temperature gradient (TGGE). In essence this approach relies upon detecting the total radioactivity associated with wild-type molecules and with those in regions of the gel which carry mutation(s). It then becomes possible to calculate the fidelity from the formula:

$$f = HeF/(b \times d)$$

where f = errors per bp incorporated/duplication;
HeF = heteroduplex fraction (the fraction of molecules carrying mutations; i.e. not running as wild type in the gel);
b = length of single-strand DNA being measured (does not include the GC anchor); and
d = number of DNA duplications.

Alternatively denaturing HPLC can be used. This is a technique increasingly important for detecting single nucleotide polymorphisms (Chapter 11) which can also be applied for detecting mutations in any DNA PCR product.

In many cases a level of misincorporation can be tolerated if the PCR DNA is to be studied as a population, for example by direct sequencing or restriction analysis of the product where randomly distributed mutations do not interfere with the predominant pattern. If you are starting with a large number of template molecules you could anticipate that any mutations would appear relatively infrequently against the predominant, wild-type sequence. On the other hand if you are starting with small numbers of templates, and a mutation occurred very early during the PCR, it would be amplified and might come to represent a major product, thus complicating the results; you will sometimes see this referred to as a 'jackpot' phenomenon. This problem is important when you want to clone individual molecules of the PCR-amplified DNA or where you are interested in very rare events, such as rare naturally occurring mutations in tissues or cells, the identification of alleles or the characterization of single DNA molecules. As mentioned above, if you clone PCR products you may need to sequence several clones to identify one that carries no errors.

PCR fidelity can be improved to reduce the number of errors; alteration of reaction conditions, including modifying the buffer and dNTP concentrations, can significantly improve the fidelity (12). Optimization of PCR conditions is covered in Chapter 4. As with any process, you will find that what you 'gain on the swings you lose on the roundabouts'. In essence the efficiency and fidelity of PCR counteract each other. If you want a large amount of product and are not too worried about the level of background mutations, then you can optimize efficiency at the expense of fidelity. Likewise if you are keen to ensure optimal fidelity you will need to be prepared to sacrifice efficiency, meaning that for a given number of cycles, you will get less product formed. In most cases you will probably want to reach a suitable compromise between these two parameters and this can be achieved by selecting the correct reaction conditions (Chapters 3 and 4). Increasingly with the use of PCR for gene cloning and manipulation, fidelity becomes an important factor and makes it more likely that you should choose a thermostable DNA polymerase that has a proofreading activity.

3.12   Thermostable proofreading DNA polymerases

Various thermostable polymerases that have a proofreading activity are available and these increase the range of experiments for which the PCR can be used, to include those where it is important to have high-fidelity copying of the template. Such enzymes, which are available in recombinant form, include: Vent$_R$® (New England Biolabs), a recombinant form of the naturally occurring enzyme available as *Tli* (Promega) isolated from *Thermococcus litoralis*; DeepVent$_R$™ (New England Biolabs) from *Pyrococcus GB-D*; *Pfu* (Stratagene, Promega) and *PfuUltra* (Stratagene) from *Pyrococcus furiosus*; *Pwo* (Boehringer, Roche) from *Pyrococcus woesei*; and UlTma™ (PE Biosystems) from *Thermotoga maritima* KOD HiFi (Novagen); Accuprime *Pfx* (Invitrogen). The proofreading ability is due to the capacity of the enzyme to discriminate between whether the nucleotide at the 3′-OH of an extending strand is correctly or incorrectly paired with the template strand. If the nucleotide is not correctly paired then the 3′→5′ activity will remove the 3′-nucleotide and the 5′→3′ synthesis activity will introduce a new nucleotide. The presence of the correct nucleotide, base paired with the template, provides a substrate for further polymerase extension.

In addition to higher fidelity, the proofreading enzymes are often more tolerant of variations in buffer conditions, and are more thermostable. The latter allows the use of higher denaturation temperatures and/or longer incubations at high temperatures, without significant loss of enzyme activity, which may be important for templates with high GC content. Their low K_m for DNA allows them to efficiently amplify from very low concentrations of DNA.

A characteristic feature of *Taq* DNA polymerase is its tendency to add a nontemplate-directed base to the 3′-end of the new DNA strand. This results in a single overhanging base that can affect the efficiency of a variety of subsequent manipulations, including cloning and mutagenesis. Proofreading enzymes generally do not result in such terminal additions, favoring blunt-ended products, although this is not always the case. For example Vent$_R$® and DeepVent$_R$® yield >95% blunt ends with the remainder having nontemplate-directed 3′-additions. For UlTma™, and probably most proofreading DNA polymerases, it specifically removes 3′-mismatched bases during a mispriming assay. Thus the enzyme will remove a discriminating 3′-nucleotide allowing it to amplify from either wild-type or mutant template. For example, allele-specific PCR (Chapter 11) relies upon the inability of a primer with a 3′-mismatch to prime on a template molecule while a primer with no 3′-mismatch will prime efficiently. In such assays a nonproofreading enzyme, like *Taq* DNA polymerase, must be used.

The propensity for proofreading enzymes to remove 3'-bases could potentially lead to problems of single-stranded primer degradation, most pronounced in the absence of dNTPs. This is perhaps most critical when using a primer with a mismatch to the template. Although the primer is present in vast excess over enzyme, the 3' 'nibbling' activity of a proof-reading enzyme could lead to loss of the mismatch. You should design the oligonucleotide so that the mismatch is not close to the 3'-end of the primer. It is also recommended that the enzyme be added as the final reagent during PCR set-up and that the dNTP concentration is not less than 200 μM for each nucleotide. If problems persist then it is possible to synthesize oligonucleotides with phosphothioate linkages that are resistant to cleavage, although this will be a more expensive option.

As discussed in Section 3.7 and Chapter 10, degenerate primers are some-times used in PCR. These contain mixtures of sequences when you do not know exactly what the sequence of the template is; in some cases inosine is incorporated at one or more positions to act as a universal base allowing pairing with any of the four normal bases. *Taq* DNA polymerase has no difficulty in extending such deoxyinosine-containing primers, but it has been shown that some, and therefore probably all, proofreading enzymes are unable to utilize such primers, presumably because they will not recognize the mispairing of inosine with A, G, C or T as it reaches the I in the primer region of the template strand. The degree of rejection of inosine-containing templates may depend upon the degree of proofreading as measured by the error rate (see *Table 3.2*). To allow labeling of the PCR product, modified nucleotides including digoxigenin-, biotin- and fluorescein-forms of dUTP can generally, though not necessarily efficiently, be incorporated by these enzymes. The modified nucleotide should partially substitute for dTTP, say by using 150 μM dTTP with 50 μM modified dUTP.

Choice of proofreading enzyme

It is important to consider the type of experiment you wish to perform and then to select the appropriate enzyme. A brief summary of properties of some proofreading enzymes is provided.

Vent® and *Tli* DNA polymerases

Vent® and *Tli* DNA polymerases are derived from the thermophilic archae-bacterium *Thermoccocus litoralis* that exists at temperatures of up to 98°C. The native enzyme is *Tli* (Promega) while Vent$_R$® (New England Biolabs) is the form cloned and expressed in *E. coli*. Vent$_R$® with a calculated half-life at 95°C of 6.7 hours is significantly more thermostable than *Taq* DNA polymerase ($t_{1/2}$ =1.6 hours at 95°C). Vent$_R$® DNA polymerase can generate long products up to 8–13 kb in length. It has 3'→5' proofreading activity giving it a higher fidelity than *Taq* polymerase by a factor of around five-fold. More than 95% of PCR product ends are blunt-ended with the remainder having a 3'-nucleotide extension. An exo⁻ form of Vent® is available.

DeepVent_R® DNA polymerase

DeepVent_R® DNA polymerase is a more thermostable enzyme than Vent_R®, also available from New England Biolabs, and was originally isolated from *Pyrococcus* GB-D that can grow at temperatures as high as 104°C. The calculated half-life at 95°C is 23 hours. DeepVent_R® is the recombinant form isolated from *E. coli*.

Pfu DNA polymerase

Pfu DNA polymerase is from the hyperthermophilic archaebacterium *Pyrococcus furiosus* and is available from various manufacturers. It has both 3'→5' exonuclease proofreading activity and 5'→3' exonuclease activity. The fidelity of the enzyme is some 12-fold higher than for *Taq* DNA polymerase. *Pfu* has also been combined with a thermostable factor in *PfuTurbo*™ DNA polymerase (Stratagene), which is reported to increase product yield, without affecting the high fidelity of the enzyme, allowing increased product from complex targets, fewer cycles of PCR and increased sensitivity of amplification from low quantities of template.

Exo⁻ versions

Versions of Vent_R® (13), DeepVent_R® and *Pfu* have been engineered to remove the 3'→5' proofreading activity and are usually designated exo⁻. You may wonder why such enzymes should be produced. In fact these are primarily intended for DNA cycle sequencing (Chapter 5). For normal PCRs these versions still display a slightly higher fidelity than *Taq* DNA polymerase, for example Vent_R® (exo⁻) has a two-fold higher fidelity, and can generate more target product than their proofreading parents. The exo⁻ versions also tend to generate blunt ends (70%) although a higher proportion of products have 3'-A extensions (30%), more like *Taq* DNA polymerase. Of course the exo⁻ enzyme is now not a proofreading enzyme!

Pwo DNA polymerase

Pwo DNA polymerase was isolated from the hyperthermophilic archaebacterium *Pyrococcus woesei* and is now available from Roche Molecular Biochemicals in recombinant form expressed in *E. coli*. The enzyme is a 90 kDa protein with a highly processive 5'→3' DNA polymerase and a 3'→5' exonuclease proofreading activity. There is no detectable 5'→3' exonuclease activity. *Pwo* has higher thermal stability than *Taq* DNA polymerase with respective half-lives at 100°C of 2 h and 5 min. *Pwo* DNA polymerase has a preference for $MgSO_4$ (2 mM) rather than $MgCl_2$. It is almost as processive as *Taq* DNA polymerase but the fidelity of the enzyme is about 10-fold higher. dUTP is a poor substrate for *Pwo* DNA polymerase and therefore it should not be used with uracil glycosylase (UNG) systems to prevent carry-over contamination (Chapter 4). *Pwo* DNA polymerase generates products suitable for blunt-end cloning.

UlTma™ DNA polymerase

UlTma™ DNA polymerase is a recombinant version of the enzyme from the hyperthermophilic Gram-negative eubacterium *Thermotoga maritima*. It has proofreading activity and is more thermostable than *Taq* DNA polymerase.

ACCUZYME DNA polymerase

ACCUZYME DNA polymerase (Bioline) possesses 5′→3′ DNA polymerase and 3′→5′ proofreading exonuclease activities, offering 47-fold higher fidelity compared to *Taq* DNA polymerase. This high fidelity is guaranteed for amplification products up to 5 kb in size. ACCUZYME DNA polymerase is supplied with buffer containing $MgSO_4$, which provides optimal final reaction conditions (2 mM Mg^{2+}) for most experiments, however for optimization purposes the $MgCl^{2+}$ concentration can be varied. As for other proofreading thermostable polymerases ACCUZYME generates blunt ends. We would recommend ACCUZYME as a solid thermostable proofreading DNA polymerase for applications which require high-fidelity and/or blunt ending.

KOD HiFi DNA Polymerase

KOD HiFi DNA Polymerase is supplied by Novagen. It is a proofreading enzyme, isolated from the extreme thermophile *Thermococcus kodakaraensis* KOD1, and displays high processivity and fidelity that allow faster, more accurate PCR amplification than with some other enzymes, such as *Pfu* DNA polymerase. For example the elongation rate is 5 times faster and processivity 10–15 times higher, than for *Pfu* DNA polymerase. KOD HiFi DNA Polymerase is also supplied in a hot-start version in which the DNA polymerase and 3′→5′ exonuclease activities are inactivated by two monoclonal antibodies. We recommend this enzyme for a wide range of applications.

3.13   *Tth* DNA polymerase has reverse transcriptase activity

Tth DNA polymerase, isolated from *Thermus thermophilus*, is a highly processive thermostable DNA polymerase making it useful for the production of long PCR products. It does not possess a proofreading activity and so has a similar fidelity to *Taq* DNA polymerase. However, it has good reverse transcriptase (RT) activity and so is useful for reverse transcriptase PCR (RT-PCR). It can transcribe mRNA transcripts into cDNA and then amplify these for applications such as analysis of gene expression levels, construction of cDNA libraries from limited amounts of tissue or detection of low levels of a transcript (Chapter 8). The RT activity is dependent upon the presence of 1–2 mM manganese ions while the DNA polymerase functions best in the presence of magnesium ions. There are two approaches to amplifying mRNAs using *Tth* polymerase. The first relies upon a reaction containing only manganese and involves an initial RT step with subsequent PCR thermal cycling to amplify cDNAs. The main drawback to this simple procedure is that the fidelity of *Tth* DNA polymerase is reduced in the presence of manganese and so cDNAs are likely to accumulate more errors. This may not be important for procedures where only the detection of an amplification product is important. In most detection assays it is likely to be the size and amount of product formed that is important, rather than whether the amplified product contains errors. However, fidelity will be important if you want to clone and study the product. Increased fidelity is possible by performing the initial RT step in the presence of manganese and then adding a chelating agent, such as EGTA, to sequester the manganese, before adding magnesium to activate the DNA polymerase activity for the PCR stages. Both approaches can be performed in a single tube. The key advantage of the *Tth* polymerase system over a standard two-enzyme approach based on a mesophilic RT and *Taq* DNA polymerase, is that *Tth* polymerase will perform both reactions at high temperature (70°C). This is an advantage for GC-rich RNA templates that often form complex intramolecular secondary structures. The high temperature at which *Tth* RT activity operates allows the template to be maintained in a denatured state, allowing efficient copying of normally difficult templates. The high temperature also ensures high specificity of primer annealing. Alternatives to *Tth* polymerase for single-tube RT-PCR that comprise enzyme mixtures can be found in Section 3.15.

3.14   Red and green polymerases and reagents

Some enzyme preparations or PCR reagent mixes include a dye that provides confidence that the enzyme or reagents have been added and correctly mixed and can assist in post-PCR analysis. For example, there is usually no need to add a loading buffer prior to agarose gel analysis of reaction products as the sample has increased density and the dyes migrate through the gel to act as a marker for migration of samples through the gel (Chapter 5). Sigma have a REDTaq™ preparation supplied at 1 U μl^{-1} that colors the reaction and is also claimed to act as a gel loading and migration marker that moves slightly faster than bromophenol blue. ABGene® supply a Red Hot® DNA polymerase that contains a red dye to monitor addition and mixing of the enzyme. They also produce a ReddyMix™ buffer containing an inert red dye and a gel loading reagent allowing monitoring of reagent addition and subsequent direct gel analysis of an aliquot of the reaction. Addition of a dye and loading reagent as part of a buffer, rather than the enzyme, should improve the qualities of loading and visualization since the concentrations are greater due to the increased volume of reagents added. Similarly Ruby Taq and Ruby Taq Master mix are available from USB Corp and contain an inert dye. GoTaq® DNA Polymerase from Promega is a *Taq* DNA Polymerase and a Green GoTaq® Reaction Buffer that can be used for routine PCR applications, including colony screening. Amplified DNA can be loaded directly onto agarose gels, without the need to add loading dye.

3.15   Polymerase mixtures: high-fidelity, long-range and RT-PCRs

Barnes (14) described a two-enzyme system for the amplification of long PCR products (5 to >20 kbp). The enzymes included predominantly *Taq* DNA polymerase and a smaller component of a proofreading DNA polymerase. As DNA synthesis proceeds, *Taq* DNA polymerase will occasionally insert the wrong nucleotide leading to a stalling of the extension process and dissociation of the *Taq* enzyme from the template. If it was the only DNA polymerase present, a subsequent reassociation event between the *Taq* DNA polymerase and mismatched 3′-end might lead to extension and thus 'fixing' of a mutation in the population of DNA molecules. By including some proof-reading enzyme, the stalled extension product can be repaired by the proofreading enzyme thereby allowing efficient extension of the product with reduced error rate (*Figure 3.6*). The length of product that can be generated generally depends upon the complexity of the starting template. The more complex the template the shorter the likely products, so most examples of very long products are from simple cloned templates. When performing long-range amplifications from genomic templates (Chapter 10) the quality of starting DNA is obviously of great importance and care should be taken to produce high quality genomic DNA. An additional benefit of an enzyme mixture is that it will work over a wider range of Mg^{2+} concentrations.

As the majority of DNA polymerase activity in such enzyme mixes tends to be due to *Taq* DNA polymerase it is expected that most products will carry a 3′-A extension. Many commercially available sources of optimized enzyme mixes are available for long-range PCR, high-fidelity PCR, amplification of GC-rich templates and RT-PCR. The following examples represent a snapshot of some available reagents, but there are continual developments being made by manufacturers.

Long-range PCR

- The *Advantage® 2* mix from Clontech comprises their AdvanTaq™ enzyme for high-yield product synthesis together with a smaller amount of a proofreading polymerase to allow increased fidelity and longer product lengths to be synthesized. The fidelity is about three times higher than the AdvanTaq™ polymerase alone. The system also incorporates a TaqStart™ antibody to allow hot starts (Chapter 4).
- The *Advantage®-Genomic* mixes (Clontech) for amplifying large genomic sequences contain *Tth* polymerase which is useful for copying long sequences, but which does not contain a proofreading activity. In addition there is a small amount of a proofreading enzyme to increase the fidelity of copying allowing up to 10 kb regions to be amplified from genomic DNA. From less complex templates it is possible to amplify longer sequences up to 40 kb.
- The *Expand™Long Template* PCR system (Roche) is a mix of *Taq* and *Pwo* DNA polymerases together with three optimized buffers for different size ranges that when used with the recommended protocols provide efficient amplifications of 0.5–12, 12–15 or >15 kb products. Even longer fragments can be generated using the Expand™ 20 kb PLUS PCR system.
- *Hot Tub™* from *Thermus 'ubiquitous'* (Amersham Pharmacia Biotech) is not a mixture of enzymes, but is a single enzyme preparation that is produced specifically for long-range PCR, and so is most appropriately listed here.

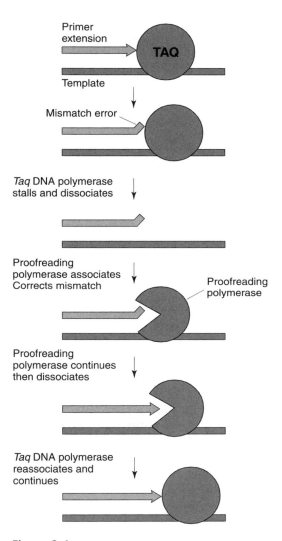

Figure 3.6

A mixture of *Taq* DNA polymerase and a proofreading enzyme allows efficient repair of errors during PCR. If *Taq* DNA polymerase introduces a mismatched nucleotide it will stall and will dissociate from the template. The error can be corrected by the proofreading DNA polymerase that will continue to synthesize DNA until it also dissociates from the template allowing *Taq* DNA polymerase to reinitiate DNA synthesis. The use of a mixture containing predominantly *Taq* DNA polymerase with a low concentration of a proofreading enzyme improves the fidelity over *Taq* DNA polymerase alone, but allows high levels of product to be synthesized.

- The *TaqPlus® Long PCR system* (Stratagene) is a combination of *Taq2000* and *Pfu* DNA polymerases allowing amplification of products up to 35 kb with two buffers supplied and extension times of only 30 s per kb.
- The *Extensor Long PCR* kit (ABGene®) is a mix of *Taq* and a proofreading DNA polymerase for 3 to >20kb products.
- The εLONGASE® *Amplification System* (Invitrogen) represents a combination of *Taq* and *Pfu* DNA polymerase giving a five-fold-enhanced fidelity over *Taq* DNA polymerase alone. It uses a single buffer system for amplifications of products 250 bp to 30 kbp.
- *KOD XL DNA Polymerase* is a blend for amplifying very long templates. This comprises KOD HiFi DNA Polymerase and a mutant form of KOD HiFi deficient in 3'→5' exonuclease activity. This preparation also allows the incorporation of dITP, dUTP and derivatized dNTPs into PCR products and produces a mixture of PCR products with blunt and 3'-dA overhangs.

High-fidelity PCR mixtures

- The *Advantage®-HF 2* PCR kit (Clontech) is also available and is designed to achieve a fidelity 25-fold higher than AdvanTaq™ alone.
- The *Expand™ High Fidelity* PCR system (Roche) comprises a mixture of *Taq* and *Pwo* DNA polymerases to provide optimized fidelity and yield during amplification of genomic fragments of less than 5 kb. The yield is two times higher than that of *Taq* DNA polymerase while the fidelity is three times higher. The Expand™ High Fidelity also provides improved sensitivity with limiting quantities of genomic DNA template.
- The *TaqPlus® Precision PCR system* (Stratagene) is a mix of *Taq2000* and *Pfu* DNA polymerases designed for amplifications up to 15 kb and with 2.7-fold higher fidelity than *Taq* DNA polymerase alone.
- *BIO-X-ACT Long DNA Polymerase* (Bioline) is a high-performance proprietary complex of enzymes and additives designed for applications requiring high processivity and high fidelity. BIO-X-ACT Long DNA Polymerase has a 17-fold increase in fidelity compared to regular *Taq* polymerase and is recommended for amplification of genomic DNA fragments between 2 kb and 20 kb. BIO-X-ACT possesses 5'→3' DNA polymerase activity and 3'→5' proofreading activity and is supplied with its own unique buffer and also with 'HiSpec Additive', which reduces smears, primer-dimers and spurious bands that are associated with difficult GC-rich and repetitive sequences.
- *Platinum® Taq DNA polymerase* (Invitrogen) is a mixture of *Pyrococcus* species GB-D polymerase, which displays 3'→5' proofreading activity, and recombinant *Taq* DNA polymerase with the Platinum *Taq* antibody to allow hot start. The mixture displays a 6-fold enhanced fidelity compared with *Taq* DNA polymerase, and allows 12 to 20 kb to be amplified following optimization.

GC-rich template PCR

- The *Advantage®-GC 2* system (Clontech) provides a polymerase mix designed to increase the efficiency of copying difficult templates, such as GC-rich regions or genomes that often prove difficult or intractable to PCR. It also includes DMSO in the buffer to disrupt base pairing of the DNA template while a reagent, GC-Melt™, is also included to assist destabilization of the template and to make AT and GC base pairs equally stable.

- The *GC-Rich PCR system* (Roche) combines *Taq* and a proofreading enzyme plus buffer additives and a resolution solution to achieve PCR of GC-rich templates.

RT-PCR

- The *C. therm. Polymerase One-Step RT-PCR System* (Roche) provides an alternative to the use of *Tth* polymerase. The Klenow fragment of DNA polymerase I from *Carboxydothermus hydrogenoformans* has a reverse transcriptase activity that functions at between 60 and 70°C thereby increasing the specificity and yield of the RT step, particularly with difficult templates. However, the enzyme does not require manganese that can lead to reduced fidelity. Rather, it is dependent upon magnesium and has a 3′→5′ proofreading activity leading to a two-fold higher fidelity than *Tth* polymerase. The subsequent PCR amplification of cDNAs is performed by the *Taq* DNA polymerase, allowing a simple one-tube approach. The *C. therm.* enzyme is also supplied separately for use in two-step RT-PCRs for GC-rich templates. In this case the reaction is performed in the presence of DMSO to assist denaturation of secondary structure and 1–2 M betaine to improve transcription of GC-rich sequences.
- The *Titan™ One Tube RT-PCR system* (Roche) combines AMV reverse transcriptase with Expand™ PCR enzyme mix of *Taq* and *Pwo* DNA polymerases to achieve RT-PCR products up to 6 kbp with higher fidelity than a *Tth* polymerase-based approach.
- *ProSTAR™ RT-PCR systems* (Stratagene) are available in HF (High Fidelity) or Ultra HF (Ultra High Fidelity) formats. The former is a combination of MMLV reverse transcriptase for first-strand cDNA synthesis and *TaqPlus® Precision* for the PCR stage. The system allows single-tube, high-yield and fidelity PCR from an mRNA template. The Ultra HF version uses again MMLV reverse transcriptase with *PfuTurbo™* DNA polymerase for the highest fidelity amplification of the cDNAs. This system is useful for cloning, sequencing and expression of products.
- The *OneStep RT-PCR kit* (Qiagen) uses a blend of two reverse transcriptases and a HotStar *Taq* DNA polymerase.

3.16   Nucleic acid templates

A wide range of DNA and RNA samples can be used as templates for PCR, including genomic DNAs, mRNAs, cDNAs, libraries, plasmid, phage, cosmid, BAC and YAC clones. In addition to preparing the template yourself there are an increasing number of commercial suppliers providing genomic DNAs, genomic libraries (in various vectors), cDNA libraries, total RNAs and poly(A)$^+$ RNAs. These are available from several animal and plant species, tissues and cell types.

The amount of template required is likely to require optimization, but generally less than a nanogram of cloned template and up to a microgram of genomic DNA is used. For high molecular weight genomic samples PCR should prove successful, but it may be useful to do a partial digestion of the DNA with rare-cutting restriction enzymes such as *Not*I or *Sfi*I. PCR can be used to amplify from single cells or even a single template molecule although it is critical in such cases to avoid contamination (Chapter 4). It is also possible to perform some PCR analyses *in situ* within a fixed sample of tissue to explore the expression of genes at the level of tissue or cells or to identify the presence of viral transcripts associated with disease (Chapter 8).

Sources of template DNA

DNA from almost any source has been successfully used in PCR. Whatever the source, make certain that you know what safety rules to apply to the handling and disposal of the starting material and/or by-products of the extraction; this may involve correct disposal of bacterial cultures, human or animal tissues, transgenic plant tissue or chemical reagents such as phenol.

In general, as with any biological system, PCR is almost certain to work best with the cleanest, purest DNA sample you can prepare. However, the inherent capacity of PCR to amplify even minute amounts of DNA means that even very crude tissue samples which have simply been lyzed, for example by boiling, will often produce a product after PCR. However, in some cases you may find that you get no amplification products from certain DNA sources, particularly those which contain significant amounts of polysaccharides, or which look an 'unusual' color. It might be that the DNA contains some contaminant that is inhibiting *Taq* DNA polymerase, and you may find that by testing a variety of dilutions of the DNA preparation, you get to a point where amplification works successfully because the concentration of contaminant has been sufficiently reduced.

Genomic DNA

Essentially any method can be used to prepare genomic DNA from bacteria, fungi, plants and animal. For the amplification of large fragments (>1000 bp) purer genomic DNA will mean better amplification and there are a number of very good commercially available DNA purification kits. However, when dealing with a large number of samples, preparation of highly purified DNA can become very time consuming. Crude sample preparations will often suffice for successful amplification of small products (200–1000 bp).

RNA isolation and first-strand cDNA preparation

There are a number of ways to isolate RNA for reverse transcriptase applications. Standard methods include the lysis of cells and tissues, and extraction of RNA followed by ethanol precipitation. Various methods can be found in most molecular biology manuals and these normally work well.

However, these extraction procedures usually include a phenol step and so are not recommended when dealing with a large number of samples. For successful reverse transcription it is important to have the highest quality RNA possible and there are many good commercial RNA isolation kits which do not include phenol extraction. Reverse transcription is covered in detail in Chapter 8.

Plasmid and bacteriophage DNA

DNA prepared by any method from rapid mini-preps to purification through CsCl gradients may be used as template for PCR experiments. For plasmid screening there is really no need to purify DNA from bacteria or phage since small parts of colonies or plaques can be picked directly into PCR reaction mixes (*Protocol 6.2*). Most molecular biology supply companies have proprietary systems for the simple, routine and reliable purification of plasmids (Chapter 6).

Pathological and forensic samples

Clearly with human tissue or blood samples you must take the appropriate precautions to prevent contamination with infective agents. DNA can usually be extracted from fresh tissues or blood samples very simply by a proteinase K digestion procedure. However, it is also possible to isolate DNA from archival samples stored either as fixed tissues on microscope slides or as museum specimens in an appropriate preservative agent (15).

Archaeological samples

DNA can be extracted and amplified from ancient biological specimens (16). The focus of such studies is often chloroplast or mitochondrial DNAs or genomic ribosomal RNA genes that occur in multiple copies per cell and therefore are likely to be better represented than the one or two copies of single-copy genes in genomic DNA. Over time the quality of the template DNA will have degenerated. Often this results in amplification of only short products of 100–200 nt (17). However, it is possible to reconstruct longer sequences by overlap analysis of many shorter segments (18,19). An alternative approach is the initial reconstruction of a longer template by recombinant PCR from purified short initial PCR products (20). Of course there is always the risk in such experiments that artefactual products may be generated that did not exist in the original sample. It would be prudent to perform parallel reactions to provide confirmation that a product was generated in independent experiments. In any studies of ancient samples it is critical to ensure appropriate controls are performed to ensure that amplification products do not arise from more recent microbial or other contamination. DNA amplification studies have been carried out on a wide range of samples such as Siberian woolly mammoth mitochondrial DNA (21), insects in amber using nuclear 18S rRNA (22,23) or mitochondrial DNA (24), leaf fossils using chloroplast DNA (25) and nuclear 18S rRNA (26) and Neanderthal DNA (18).

Amount of template DNA required

The amount of template DNA required for a PCR varies according to the application and the source of the template. As a comparative guide for 3×10^5 template molecules, which is a reasonable number to start with in a PCR, you would need:

- 1 µg human genomic DNA
- 10 ng yeast genomic DNA
- 1 ng *E. coli* genomic DNA
- 1–2 pg of plasmid or M13 DNA
- 20 pg of bacteriophage λ DNA

Increasing the number of template molecules is recommended in some applications to reduce the number of amplification cycles, for example in some gene splicing and mutagenesis procedures (Chapter 7).

3.17   Mineral oil

In thermal cyclers that do not have a heated lid (Section 3.19) it is necessary to overlay the PCR reactants with a barrier that will prevent thermal evaporation. A high quality light mineral oil, such as that available from Sigma or Aldrich, has traditionally been used. An alternative is silicone oil, which creates a flatter and clearer phase transition interface and which is available in a dropper bottle format from Invitrogen.

3.18   Plasticware and disposables

A number of manufacturers supply disposable items for PCRs. These include, 0.5 and 0.2 ml polypropylene tubes and 96-well microtiter plates, glass slides, glass and plastic capillaries and pipette tips. Tubes and microtiter plates are designed to give uniform contact with reaction blocks and are available in thin-wall format for rapid heat conduction between the block and reactants to improve cycling efficiency. Tubes are available in strips. Caps are also available in strips while seals for microtiter plates are available either as strips of caps or self-adhesive or heat seal membranes; the latter require access to an appropriate heat sealer unit. Pipette tips are available that prevent contamination due to aerosols by the use of filters, or for positive displacement pipettes, and are discussed further in Chapter 4. When using 96-well microtiter plates multi-channel pipettes are recommended as this reduces pipetting error and is less time consuming. For high throughput studies it can be useful to make use of a liquid handling robot if possible. Racks that accept a mixture of tube sizes can be useful for PCR set-up, since it is unlikely that your reagents will all be stored in the same size tubes as your PCR reaction tubes. The type of tube you use will affect the temperature profile of your PCR. As an example *Figure 3.7* shows a comparison of a set thermal cycle of 94°C, 1 min; 55°C, 1 min; 72°C, 1 min, on an instrument that monitors block, rather than reaction temperature. The reaction temperatures were monitored in two different types of tubes, a standard 0.5 ml polypropylene tube and a thin-walled PCR tube. The traces show that the normal tube never reaches the preset temperature before the

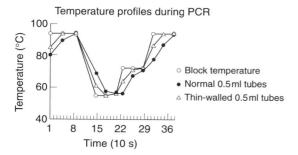

Figure 3.7

The effect of altering tube type on the temperature profile of a PCR reaction. The set profile is shown together with actual reaction temperature monitored in either a standard or thin-walled reaction tube on the same instrument. The results demonstrate the significant difference in reaction conditions.

transition to the next temperature. Thus for example the reactions are being carried out at an annealing temperature above that set. By contrast the thin-walled tube reaches the preset temperatures by around 30 s allowing a 30 s hold at the set temperature. It is important to understand how parameters such as reaction tubes can affect the kinetics of the PCR to ensure reproducibility in your experiments.

3.19   Automation of PCR and thermal cyclers

The early instruments for PCR were rather crude and generally consisted of a robotic arm used to transfer a rack of reaction tubes between the water baths held at different temperatures. Such devices were not particularly efficient and were generally replaced by programmable heating blocks, commonly called thermal cyclers. These instruments are based on mechanisms for rapid heating and cooling of the reaction tubes to a predetermined temperature and holding the temperature for a defined time before moving to the next temperature. Various mechanisms for heating and cooling exist and the most common are heating by electrical resistance with either fluid or semiconductor (Peltier device) cooling, or heating using a powerful lamp or heating coil with air cooling. Many manufacturers supply instruments for PCR and these range from the simple through to extremely sophisticated. All tend to provide simple-to-understand programming capacity that allows the user to control the cycling parameters, rate of heating and cooling and the number of cycles. Some instruments have a single control block that can operate satellite blocks. Others have gradient blocks that can be set to a range of temperatures while the most sophisticated also have real-time systems for monitoring product accumulation.

Patents, held by Hoffman-LaRoche, were originally licensed only to PE-Biosystems, for the sale of thermal cyclers. Now, most manufacturers supply appropriately licensed instruments and these are generally very user-friendly requiring only simple menu-driven programming. Such instruments will cycle through a series of preset temperatures a predefined number of times. Some instruments allow the rate of heating and cooling (s °C^{-1}), known as ramping, to be regulated which may be important for certain experiments, for example using long primers. They may also provide additional features such as preprogrammed pauses to allow sampling of the reaction mix or addition of further reactants after a defined number of cycles. Some will refrigerate the samples at 4–6°C on completion. This facility can be useful, especially when performing overnight reactions when the samples might be sitting for several hours after the completion of the PCR.

Heating

Most instruments rely upon a heating element in contact with metal block suitably modified to hold the reaction tubes or microtiter plates. The rates of heating are usually around 1–2°C s^{-1}. What should concern you most is the uniformity of heating across the block. It is all very well the tubes in the middle of the block reaching the appropriate temperature, but those at the outsides should also reach this same temperature, at the same time. This does not always happen and is a feature worth checking, preferably for yourself by getting an instrument on a trial basis. There are experimental procedures where such minor temperature variation can have significant effects. Examples are the quantitation of PCR products to provide a measure of initial template concentration, or RAPD-PCR where annealing of the short primers is often highly sensitive to temperature. Air cyclers do not have a block but instead use rapid heating of air surrounding reaction capillaries.

Cooling

When it comes to cooling the block there are a range of systems, for example the Perkin-Elmer instruments use a conventional refrigerant, others use fans or solid-state Peltier devices while others use water-based cooling. The type of cooling system is not insignificant since to achieve maximal efficiency you require rapid, uniform cooling, and most systems achieve this. The other consideration is the lifetime of the instrument, or perhaps more accurately that of its components. Early Peltier devices were notorious for failing, particularly if frequently used to cool post-PCR samples to 4°C, and were quite expensive to replace. Advances in technology have improved performance and reliability.

Temperature monitoring

It is important to understand the mechanism that the manufacturer employs for controlling the cycle parameters as these are not the same on all instruments and can lead to failure when a protocol developed on one instrument is exported to a different type of instrument. An important criterion is reproducibility both in attaining and maintaining temperatures for the desired times. Some instruments monitor block temperature and use this to regulate the time of each step while others use a reaction temperature measured by a thermocouple in a tube placed in the block. As it will take longer for the reaction solution than the block to reach the programmed temperature this means that alterations need to be made if identical reactions are to be carried out on such different instruments. This is also true when moving from a block-based instrument to an air cycler (see below) which is based on capillaries exposed to air that is rapidly heated and cooled. Such instruments tend to reach their programmed temperature much more quickly than block-based instruments and therefore reaction cycles are generally very short. It is also important that the temperature of the reactants in the reaction vessel do not over- or undershoot the desired temperatures as this would prevent controlled reaction conditions.

Gradient blocks

Instruments are now increasingly available that allow rapid optimization of the annealing temperature step of PCRs. The ability to use a single block to simultaneously test 8 or 12 predetermined temperatures during optimization of PCR conditions is a significant advance. For day-to-day experiments a gradient instrument should allow different sections of the block to operate at different temperatures allowing you to perform independent experiments, requiring different annealing temperatures, simultaneously. If you are considering purchasing a new instrument and will be undertaking a range of experimental studies, rather than routine established testing, it is recommended that you seriously consider an instrument with a gradient capacity. One thing to bear in mind is that due to the varying mechanisms employed by different instrument manufacturers it is often difficult to dissect out the exact conditions of incubation time and temperature. This means it will not always be possible to optimize conditions on a gradient instrument and then to transfer the protocol to a nongradient instrument, unless it is from the same manufacturer. Instruments are being developed that provide very accurate readout of temperature dynamics that should improve the facility to transfer protocols more reliably.

Heated lids

Most manufacturers produce thermal cyclers with heated lids. If you have an instrument without a heated lid you will need to add a drop of light mineral oil to provide a thermal barrier preventing evaporation of the reaction solution. The heated lid fits tightly against the lids of the tubes ensures good thermal equilibration and overcomes the evaporation problem.

Air cyclers

Thermal cyclers are now available that do not rely upon a thermal block system. Air cyclers use a heat source such as a bulb or heating coil to direct hot air into a chamber into which are inserted sealed capillaries containing the PCR reaction mix, usually in a volume of 10–20 µl. The cooling steps are achieved by a fan, which extracts hot air and allows entry of cold air to the chamber to achieve the required annealing temperature. The rapidity of the air-mediated temperature changes and the high surface-area-to-volume ratio of the capillaries lead to very rapid alteration of the temperature of the PCR mix and therefore cycling times are significantly reduced. A typical PCR that may take 2 hours in a standard instrument can be completed in as little as 30 min in an air cycler.

Monitoring product accumulation and PCR kinetics

Systems are becoming increasingly available both for performing PCR and for monitoring the accumulation of product directly in real time after each cycle. These systems are based on fluorescent detection of double-stranded DNA or specific target sequences by measurement of fluorescence resonance energy transfer (FRET). These assay systems are discussed in greater detail in Chapter 9.

Examples of PCR instruments

The choice of instrument for your laboratory should be made on the basis of a list of your requirements and ideally the ability to test different instruments with your own samples. Many suppliers will provide a demonstration instrument for you to test, either during a more formal demonstration, or ideally by leaving it with you for a few days. It is useful to be able to directly compare instruments with identical samples to compare the quality and reproducibility of PCRs. It is a good idea to check out several competing instruments within your budget range. The better ones are obviously likely to be the most expensive, but it is not worth buying one because it is cheap (relatively speaking) if it will not do the job properly. You are better to try to come to some deal with whoever holds the purse strings, and with the suppliers, in order to get the best machine you can. The following section highlights some types of instruments that have features worthy of particular note and gives the relevant websites for further information.

Applied Biosystems instruments

Applied Biosystems have a range of thermal cyclers including the GeneAmp® PCR system 9700, the Applied Biosystems 2720 thermal cycler and the Applied Biosystems 9800 Fast thermal cycler. The instruments are based on a Peltier system and are recognized as being manufactured to the highest standards. The 9700 system comes with various reaction sample block modules including: (i) a 60-well 0.5 ml sample block; (ii) a 96-well 0.2 ml sample block; (iii) a 96-well reaction plate block, and (iv) two 384-well reaction plates block. The 2720 system is a personal-sized thermal cycler and can be used with standard MicroAmp® eight-strip reaction tubes or a MicroAmp® 96-well plate. The 9800 Fast system is more rapid than the 9700 and 2720 systems and uses a 96-well plate. Further detailed information can be found at http://www.appliedbiosystems.com/.

Thermo Hybaid instruments

Thermo Hybaid instruments encapsulate many of the features described earlier in this Section. There are gradient and nongradient versions with heated lids and in-reaction tube temperature monitoring. It is possible to program increases or decreases (touchdown PCR, Chapter 4) in temperature at each cycle, which can be useful for certain applications such as long-range PCR. For certain robust applications such as colony screening (Chapter 6) it is possible to reduce the times required for denaturation, annealing and extension due to the rapid heating rates. The PCR Sprint Thermal cycler is ideal for relatively small amounts of PCRs on a regular basis. The PCR Sprint has two choices of sample format; the 0.2 ml block option with 24 tubes and the 0.5 ml block option with space for 20 tubes. The blocks are also interchangeable, so no need to buy a new thermocycler if your requirements change. The P×E Thermal Cycler (P×E) is a licensed 96-well PCR machine similar to the PCR Sprint. If more advanced amplification parameters are needed the P×2 Thermal Cycler is a good choice designed to offer performance, accuracy and simplicity. The P×2 offers five control modes for precision temperature control, including active tube control for in-sample measurement. In addition, six interchangeable blocks are available including the latest gradient cycling format. Further detailed information can be found at http://www.thermo.com/.

Eppendorf Gradient Mastercycler

In addition to the flexibility, reproducibility and ease of programming expected in a modern gradient thermal cycler this instrument also has a novel universal block able to accommodate 96 0.2 ml tubes, 77 0.5 ml tubes (or a mixture of 0.2 and 0.5 ml) or a 96-well plate. Further detailed information can be found at http://www.eppendorf.com.

RoboCycler® temperature cyclers

Supplied by Stratagene these are a modern variation on the original theme of having water baths at preset temperature and transferring tubes between the baths to perform the PCR cycle. Robocylers have four temperature blocks, one a fixed cooling block while the other three are user-programmable for the denaturation, annealing and extension steps. A robotic arm rapidly (< 3 s) transfers tubes from one block to the next thereby overcoming the normal requirement on single-block instruments for ramping the block between the appropriate temperature. The cycling time is therefore reduced allowing more rapid completion of PCRs. The instruments have either a 40-well or a 96-well format. Gradient temperature cyclers are also available that allow simultaneous testing of 8 (40-well format) or 12 (96-well format) annealing temperatures. As expected well-to-well uniformity is good with ± 0.1°C for the standard instruments and 0.25°C for the gradient version. Further detailed information can be found at http://www.stratagene.com.

MJ Research instruments

MJ thermal cyclers pioneered the use of Peltier-effect heating and cooling that is now standard in thermal cyclers and MJ is the only cycler manufacturer to make its own Peltier modules. The introduction of the DNA Engine revolutionized thermal cycling and has become a platform for newer features and upgrades. The DNA Engine has fantastic thermal uniformity and also interchangeable blocks and is often looked upon as the 'workhorse' of thermal cyclers.

In terms of real-time RT-PCR (Chapter 9) the MJ product line offers a range of instrumentation for both single- and multi-color fluorescence detection. The Opticon™ real-time PCR detector employs an array of 96 blue light-emitting diodes (LEDs) to excite fluorescent dyes. LEDs provide a long-lived excitation source of uniform performance and each LED beam is focused onto its corresponding well to ensure minimal crosstalk and light scattering. Emitted fluorescence is detected by a photomultiplier tube (PMT) that detects and amplifies the fluorescence signal and together with a bandpass-filter detection system the specific signal is separated from the background. This results in exceptional sensitivity and the ability to accurately quantify template concentrations over a broad dynamic range. The Opticon™ 2 system has the same features as the Opticon system but has the capability of two-color detection. Sample wells are illuminated by an array of 96 blue-green LEDs but detected by two PMTs. The first PMT detects wavelengths between 523 and 543 nm (suitable for detecting SYBR Green I and FAM-labeled probes; Chapter 9), and the second PMT detects wavelengths from 540 to 700 nm (suitable for detecting fluorophores, such as HEX, TET, TAMRA; Chapter 9). The Chromo4 detector system is particularly attractive since it easily converts any DNA Engine into a real-time RT-PCR system by simply swapping with any of the interchangeable sample blocks. The flexibility of the Chromo4 allows it to keep pace with such a rapidly evolving field. Further detailed information can be found at http://www.mjr.com.

LightCycler™ instrument

This instrument (Roche) is an air cycler allowing rapid PCR combined with real-time amplicon detection (Chapter 9). An additional feature allowing melting curve analysis also makes the instrument suitable for rapid product analysis and mutation detection studies. The instrument has the benefits of an air cycler in facilitating rapid cycling allowing 30–40 cycles to be performed in 20–30 min and accepts up to 32 capillaries in a carousel. The glass capillary also facilitates fluorescent signal detection by acting as an optical element. The carousel is rotated to position each capillary above the fluorescence monitoring unit which comprises a light-emitting diode light source (470 nm emission) and three band-pass filters (530, 670 and 710 nm) allowing simultaneous detection of three fluorophores. Fluorescence output is displayed in real time. Lightcyclers that use 96 and 384 well plate format are also available. Further detailed information can be found at http://www.roche-applied-science.com/.

In situ PCR instruments

Instruments are also available for performing *in situ* PCR experiments based on microscope slides. Essentially such instruments are rather similar to standard thermal cyclers except that they must have a special block capable of accepting slides that have been appropriately sealed to ensure reagents remain in contact with the samples during cycling. Eppendorf has developed an *in situ* adapter that fits on universal blocks of the Eppendorf Mastercycler thermal cyclers. This adapter, with room for up to four standard glass slides, provides perfect control of sample temperature. The adapter also contains a removable humidity chamber. Further detailed information can be found at http://www.eppendorf.com.

Further reading

Bebenek K, Kunkel TA (1995) Analyzing fidelity of DNA polymerases. *Methods Enzymol* **262**: 217–232.
McPherson MJ, Jones KM, Gurr S-J (1991) PCR with highly degenerate primers. In McPherson MJ, Quirke P, Taylor GR (eds) *PCR1: A Practical Approach*, pp. 171–186. Oxford University Press, Oxford, UK.

References

1. Eckert KA, Kunkel TA (1995) The fidelity of DNA polymerases in the polymerase chain reaction. In McPherson MJ, Quirke P, Taylor GR (eds) *PCR2: A Practical Approach*, pp. 225–244. Oxford University Press, Oxford, UK.
2. Suggs SV, Wallace RB, Hirose T, Kawashima EH, Itakura K (1981) Use of synthetic oligonucleotides as hybridization probes. 3. Isolation of cloned cDBA sequences for human β-2-microglobulin. *Proc Natl Acad Sci USA* **78**: 6613–6617.
3. Sambrook J, Fritsch EF, Maniatis T (1989) *Molecular Cloning: A Laboratory Manual* (2nd edition). Cold Spring Harbor Laboratory Press, Cold Spring Harbor, NY.
4. Wu DY, Ugozzoli L, Pal BK, Qian J, Wallace RB (1991) The effect of temperature and oligonucleotide primer length on the specificity and efficiency of amplification by the polymerase chain reaction. *DNA Cell Biol* **10**: 233–238.
5. Nichols R, Andrews PC, Zhang P, Bergstrom DE (1994) A universal nucleoside for use at ambiguous sites in DNA primers. *Nature* **369**: 492–493.
6. Loakes D, Brown DM (1994) 5-Nitroindole as an universal base analog. *Nucleic Acids Res* **22**: 4039–4043.
7. Brock TD, Freeze H (1969) *Thermus aquaticus* gen. n. and sp. n., a nonsporulating extreme thermophile. *J Bacteriol* **98**: 289–297.
8. Kunkel TA, Soni A (1988) Exonucleolytic proofreading enhances the fidelity of DNA synthesis by chick embryo DNA polymerase-λ. *J Biol Chem* **263**: 4450–4459.
9. Bebenek K, Joyce CM, Fitzgerald MP, Kunkel TA (1990) The fidelity of DNA synthesis catalyzed by derivatives of *Escherichia coli* DNA polymerase I. *J Biol Chem* **265**: 13878–13887.
10. Kunkel TA (1985) The mutational specificity of DNA polymerase-b during *in vitro* DNA synthesis – production of frameshift, base substitution and deletion mutations. *J Biol Chem* **260**: 5787–5796.
11. Fodde R, Losekoot M (1994) Mutation detection by denaturing gradient gel electrophoresis (DGGE). *Hum Mut* **3**: 83–94.
12. Eckert KA, Kunkel TA (1991) The fidelity of DNA polymerases used in PCRs. In McPherson MJ, Quirke P, Taylor GR (eds) *PCR1: A Practical Approach*, pp. 225–244. Oxford University Press, Oxford, UK.

13. Kong H, Kucera RB, Jack WE (1993) Characterization of a DNA polymerase from the hyperthermophile archaea *Thermococcus litoralis*. Vent DNA polymerase, steady state kinetics, thermal stability, processivity, strand displacement, and exonuclease activities. *J Biol Chem* **268**: 1965–1975.

14. Barnes WM (1994) PCR amplification of up to 35-kb DNA with high fidelity and high yield from lambda bacteriophage templates. *Proc Natl Acad Sci USA* **91**: 2216–2220.

15. Jackson DP, Hayden JD, Quirke P (1991) Extraction of nucleic acids from fresh and archival material. In McPherson MJ, Quirke P, Taylor GR (eds) *PCR1: A Practical Approach*, pp. 29–50. Oxford University Press, Oxford, UK.

16. Hoss M, Pääbo S (1993) DNA extraction from Pleistocene bones by silica-based purification method. *Nucleic Acids Res* **21**: 3913–3914.

17. Pääbo S (1989) Ancient DNA: extraction, characterization, molecular cloning and enzymatic amplification. *Proc Natl Acad Sci USA* **86**: 1939–1934.

18. Krings M, Stone R, Schmitz H, Krainitzki M, Stoneking M, Pääbo S (1997) Neanderthal DNA sequences and the origin of modern humans. *Cell* **90**: 19–30.

19. Stone AC, Stoneking, M (1998) mtDNA analysis of a prehistoric Oneota population: implications for the peopling of the new world. *Am J Hum Genet* **62**: 1153–1170.

20. Nasidze I, Stoneking M (1999) Construction of larger-size sequencing templates from degraded DNA. *BioTechniques* **27**: 480–481.

21. Hoss M, Pääbo S, Vereshchagin NK (1994) Mammoth DNA sequences. *Nature* **370**: 333.

22. Cano RJ, Poinar HN (1993) Rapid isolation of DNA from fossil and museum specimens suitable for PCR. *BioTechniques* **15**: 432.

23. Cano RJ, Poinar HN, Pieniazek NJ, Acra A, Poinar GO (1993) Amplification and sequencing of DNA from a 120–135 million year old weevil. *Nature* **363**: 536–538.

24. De Salle R, Gatesy J, Wheeler W, Grimaldi D (1992) DNA sequences from a fossil termite in oligomiocene amber and their phylogenetic implications. *Science* **257**: 1933–1936.

25. Golenberg EM, Giannasi DE, Clegg MT, Smiley CJ, Durbin M, Henderson D, Zurawski G (1990) Chloroplast DNA sequences from a miocene magnolia species. *Nature* **344**: 656–658.

26. Poinar HN, Cano RJ, Poinar GO (1993) DNA from an extinct plant. *Nature* **363**: 677.

Protocol 3.1 Phosphorylation of the 5′-end of an oligonucleotide

EQUIPMENT

1.5 ml microcentrifuge tube

Adjustable heating block or water bath

Ice bucket

MATERIALS AND REAGENTS

Polynucleotide kinase and accompanying buffer

10 mM ATP solution

DNA template, i.e. an oligonucleotide primer

Sterile double distilled water or equivalent

1. Mix the following reagents in a microcentrifuge tube:
   - 2 μl 10 × kinase buffer (supplied with enzyme);
   - 5 μl primer (20 pmol μl^{-1});
   - 2 μl 10 mM ATP;
   - 1 μl polynucleotide kinase (5–10 units);
   - 10 μl water.

2. Incubate at 37°C for 30 min.

3. Inactivate the kinase by heating to 90°C for 5 min.

4. Place on ice if to be used immediately, or store at –20°C for later use.

Optimization of PCR

4

4.1 Introduction

Depending on the success of your PCR amplification it may be necessary to optimize the conditions. This Chapter deals with various aspects of PCR optimization including reagents, temperatures, enhancers and preventing contamination. There is also a troubleshooting guide that will hopefully help you identify the source of any problems.

Control reactions

It is important to perform control reactions in parallel with the test samples to indicate whether any specificity (Section 4.2) or contamination (Section 4.5) problems exist. At least two controls are essential, a reaction containing no DNA and one containing no primers. You should think of the control reactions as being just as important as your test samples and there are times when you may wish to include more controls. For example, if you are beginning to work with a new pair of primers, it is a good idea to include controls containing single primers. In this way you can see whether any products are generated from either of the primers alone, rather than by the two working in combination.

4.2 Improving specificity of PCR

Primer pairs do not all work under the same reaction conditions. In some cases under 'standard' conditions one pair of primers will work very efficiently and give rise to a unique product in large amounts. At the same time, and under identical reaction conditions with the same template DNA, another primer pair will give rise to either no product, or to a complex pattern of extraneous products. Even more perplexing, sometimes you can take one primer (primer A) that you know works well with primer B, but when you use it in combination with a new primer C, the PCR fails.

The rules governing the operating characteristics of a primer pair are not defined. In essence the strategy that is usually followed for a new primer pair is to start with 'standard' amplification conditions (such as *Protocol 2.1*). If the PCR is not optimal then one of the reaction parameters should be changed to increase or decrease the stringency of the reaction conditions appropriately. If no bands are detected then the stringency may be too high whilst if several bands are seen then the stringency should be increased. How do you set about this optimization task? Are there standard rules or is it an empirical 'hit-and-miss' process? The answer lies somewhere between these two extremes. The most important parameters that will influence reaction specificity are the annealing temperature, the cycling regime and the buffer composition.

High specificity in PCR is favored by:

- optimal concentration of Mg^{2+}, other ions, primers, dNTPs and DNA polymerase;
- efficient denaturation, high annealing temperatures and fast ramping rates;
- touchdown PCR;
- hot-start PCR;
- booster PCR;
- limiting the number of cycles and their length;
- thermal cycler efficiency;
- PCR additives;
- template quality (Section 4.3);
- nested PCR (Section 4.4).

Magnesium ions

As you already know (Chapter 3) the concentration of magnesium ions (Mg^{2+}) is critical. It exists as dNTP-Mg^{2+} complexes that interact with the sugar-phosphate backbone of nucleic acids and influence the activity of the DNA polymerase. So, altering the concentration of $MgCl_2$ can lead to one primer/template pair behaving significantly differently from another under identical conditions. The usual strategy for assaying the effect of Mg^{2+} ion concentration is to adjust the standard buffer so that the $MgCl_2$ concentration varies between 0.5 and 5 mM, usually in steps of 0.5 or 1 mM. Often one concentration will show a significantly improved PCR product pattern. Remember also that you will change the Mg^{2+} concentration if you alter the concentration of dNTPs in the reaction (Chapter 3). If the Mg^{2+} concentration is too low then yields are likely to be poor while excess Mg^{2+} can reduce the fidelity of *Taq* DNA polymerase and lead to amplification of nonspecific products.

Other ions

A study by Blanchard *et al.* (1) has gone some way towards standardizing buffer conditions and variations that may lead to rapid optimization of PCRs. They used a set of buffers called TNK that contain Tris-HCl (pH 8.3), ammonium chloride (NH_4Cl), potassium chloride (KCl) and magnesium chloride ($MgCl_2$) and analyzed the effects of varying the concentrations of these buffer components. Interestingly, they found that potassium and ammonium ions, which in many biological systems behave interchangeably, gave opposite effects in the PCR. Increasing KCl leads to a reduced stringency by affecting the melting characteristics of DNA by neutralizing the negative charge of the phosphate groups of the backbone so that the hydrogen bonding between bases becomes more important. Indeed at very high KCl concentrations (>0.2 M) this stabilizing effect becomes so pronounced that the DNA strands will not denature at 94°C and therefore no PCR can occur.

DNA polymerase concentration

If you have no product band(s) or weak band(s) you may have too little DNA polymerase. Different versions of thermostable DNA polymerase can be supplied at a variety of specific activities and concentrations. To check whether this is the problem you can perform a titration with varying amounts of DNA polymerase. Thermostable proofreading DNA polymerases can have lower processivity than *Taq* DNA polymerase, affecting yield, and so more enzyme may be needed for successful amplification. It is also important to remember that thermostable polymerases will become inactivated at high temperatures and this can lead to reduced levels of product. So, try to limit the time the enzyme spends above 90°C by using a short denaturation time at 94°C, say 15 s, or a lower denaturation temperature of 92°C rather than the 94°C recommended in many protocols.

Temperatures

Denaturation

It is important that the template is efficiently denatured in order to provide single-stranded templates for PCR. This is achieved during the initial, usually 5 min denaturation phase when the sample is heated to around 94°C. If this step is inefficient then partially denatured duplex molecules will rapidly reassociate to prevent efficient primer annealing and DNA extension. For GC-rich templates it may be necessary to increase the temperature of this step to, for example, 96°C. However, it is not clear that such an extended time is required for many applications. With the exception of GC-rich templates, or where you are using a hot-start enzyme, the time could probably be reduced to 1 or 2 min. This would have the additional benefit of extending the useful life of the thermostable DNA polymerase. At the start of each cycle there is a shorter denaturation step that should denature the PCR products for subsequent reaction. While many protocols use a 94°C step here, for many templates it may be sufficient to use a temperature of 90–92°C, although GC-rich ones may require a higher temperature. It is useful to try to use the lowest effective temperature for the shortest effective time in order to retain the highest DNA polymerase activity in the reaction.

Annealing

The success of a PCR relies heavily on the specificity with which a primer anneals only to its target (and not nontarget) sequence so it is important to optimize this molecular interaction. Whether a primer can anneal only to its perfect complement, or also to sequences that have one or more mismatches to the primer, depends critically upon the annealing temperature. In general the higher the annealing temperature the more specific the annealing of the primer to its perfect matched template and so the greater the likelihood of only target sequence amplification. The lower the temperature, the more mismatches between template and primer can be tolerated leading to increased amplification of nontarget sequences. In practice it is often feasible to start at a temperature such as 55°C and assess the success of your PCR. If there is poor recovery of product and a high background of nonspecific products then empirical determination of an optimal annealing temperature may be necessary, coupled with optimization of the MgCl$_2$ concentration (see above). It is also worth checking that the time for annealing is not too long. Generally about 30–60 s is reported in methods and the shorter the better. Since the polymerase will have some activity at the annealing temperature, the longer you hold the reaction at this temperature the increased risk there is of amplification of nonspecific products.

Adjusting the annealing temperature step can alter the specificity of pairing between template and primer. If there is no product, the temperature may be too high and can be reduced, for example from 55°C to 50°C in the first instance. At the new temperature the primers may be more efficient. If there are products in control lanes where only one primer is present this indicates that the single primer is annealing to more than one region of the template and generating products. In this case you should increase the annealing temperature. As described in Chapter 3 thermal cyclers are now widely available that have a gradient block, allowing the simultaneous determination of optimal annealing temperature profiles in one reaction. By aliquoting a reaction premix into a series of tubes, the only variable should be the annealing temperature applied by the gradient block. If you do not have access to such an instrument an appropriate way to optimize primer/template annealing is to test by setting up PCR reactions and carrying out a series of experiments with 2–5°C adjustments of the annealing temperature.

There are examples of two-step PCR where the primers can anneal to the template at 72°C thereby allowing cycling between the denaturation temperature and the extension temperature. Two-step PCRs are often performed for difficult PCRs such as amplification of large fragments from genomic DNA.

Several approaches that rely upon temperature-based control of primer annealing have proven useful in improving the specificity of primer annealing and therefore of amplification of the desired product. These are considered in the next two Sections.

Touchdown PCR

Touchdown PCR starts initially with an annealing temperature higher than the T_m of the primers and then at each of the earlier cycles of the PCR the annealing temperature is lowered gradually to below the T_m. This ensures that only specific annealing of the primers to their correct target sequence takes place before any nonspecific annealing events. A good rule of thumb, described by Don *et al.* (2), when using primers about 20 nucleotides in length, is to reduce the annealing temperature by 1°C every 2 cycles moving from 65°C to 55°C over the first 20 cycles. The reaction should then be completed by another 10 cycles at a 55°C annealing temperature. Since the first products to be made are specific products, this increases the concentration of true target sequences in the early stages of the PCR thereby enhancing the accumulation of true product as the amplification continues at a less specific annealing temperature.

Hot-start PCR

Even if you take great care in designing primers and in determining the most appropriate annealing conditions, specificity problems can arise even before the first cycle of PCR. How is this possible? Consider what happens when the various reagents and template are added to the PCR tube at room temperature or on ice and then placed in a thermal cycler to start the reaction. The tube may be left for some time before being placed in the thermal cycler. It is then heated up to 95°C in order to denature the template. However, during the time it is standing at or below room temperature, until it reaches a temperature of around 65 to 70°C, non-specific primer/template and primer/primer annealing events may occur to provide substrates for the DNA polymerase. Any products formed in this manner will be templates for subsequent amplification resulting in non-specific products and/or primer-dimers. The simplest way of avoiding such spurious priming events by enhancing correct primer annealing is by the use of a 'hot-start' procedure (3–5), which relies upon the physical separation of reagents until a high temperature has been reached. One or more reactant is omitted until the temperature of the reaction is above 70°C. The final reactant(s) can then be added to allow the reaction to proceed.

There are various strategies for performing hot-start PCR; the cheapest procedure is to set up the complete reactions without the DNA polymerase and incubate the tubes in the thermal cycler to complete the initial denaturation step at >90°C. Then, while holding the tubes at a temperature above 70°C, the appropriate amount of DNA polymerase can be pipetted into the reaction. But remember if you are using mineral oil you must put the pipette tip through the mineral oil layer first, so that the polymerase is introduced into the reaction rather than floating around on top of the oil. This approach can be used in a research laboratory where relatively small numbers of reactions are being performed. However, it is not suitable for processing large numbers of samples due to:

- the time involved in making additions of enzyme to individual tubes;
- the 'loss-of-concentration' phenomenon leading to failure to add enzyme to one or more tubes; and
- the opportunity for contamination due to the need to open the tubes (Section 5).

Various commercial reagents are now available to facilitate hot starts and such products are recommended for routine hot-start applications. Some examples are given below.

Inactive DNA polymerase

Probably the most common approach used for hot start is DNA polymerase whose polymerase and in some cases 3'→5' exonuclease activity has been inhibited by the physical binding of inactivating monoclonal antibodies that prevent it reacting with substrates (*Figure 4.1*). This allows all the reaction components to be mixed together in the absence of any polymerization. When the reaction reaches a high temperature the anti-

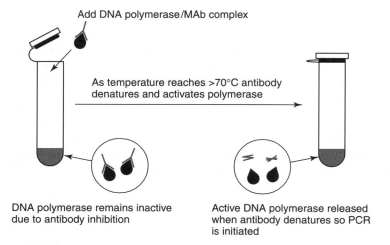

Add DNA polymerase/MAb complex

As temperature reaches >70°C antibody denatures and activates polymerase

DNA polymerase remains inactive due to antibody inhibition

Active DNA polymerase released when antibody denatures so PCR is initiated

Figure 4.1

Hot-start PCR using a thermostable DNA polymerase-inactivating antibody complex. The antibody sterically blocks the enzyme active site preventing the DNA polymerase from functioning until the antibody is denatured at high temperature.

body(s) denatures thereby releasing the thermostable DNA polymerase in an active form, allowing polymerization (*Figure 4.1*). There are many DNA polymerases of this type sold by a range of companies. These require sufficient time during the initial denaturing step to inactivate the antibodies, but usually this is achieved by a 5 min soak at 94°C. Even if it is not fully activated during this step, it will activate during thermal cycling at each denaturation step during the early cycles of a PCR.

Hot-start procedures are most useful when low concentrations of a complex template, such as genomic DNA, are being used. However, artefactual amplifications can occur in any reaction and it is generally recommended that all PCRs should be performed under a hot-start procedure.

Wax beads

The principle of wax beads, such as Ampliwax (Applied Biosystems) or DyNAwax (Finnzymes), is to physically separate some reaction components until the entire reaction has reached a high temperature, where mispriming events will not occur. As illustrated in *Figure 4.2*, some reactants, such as buffer, dNTPs, primers, template DNA and Mg^{2+}, are placed in the reaction tube. A wax bead is added and the reaction incubated in the thermal cycler at 75–80°C for 5–10 min to melt the wax. The tube is then cooled to below 35°C to allow the wax to solidify and form a barrier layer above the initial reactants. The thermostable DNA polymerase can then be pipetted on to the wax layer and the PCR cycling started. As the temperature rises the wax melts and the enzyme becomes mixed with the other reactants to initiate the PCR while the wax rises to the surface. The wax layer has the added

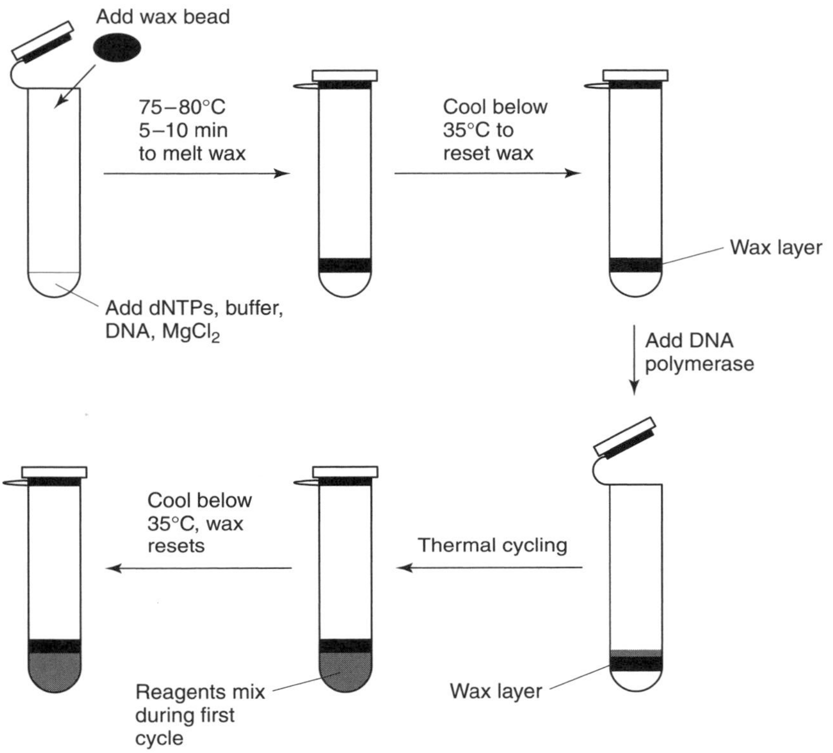

Figure 4.2

Hot-start procedure using wax beads. Some reagents are added to the tube before a wax bead is added and melted. Once the wax has solidified to form a barrier over the reactants, the missing reagents are pipetted onto the wax layer. When the wax layer melts during the first heating step of the PCR all the reagents become mixed and the reaction is initiated.

benefit of providing a barrier against evaporation during thermal cycling, in place of mineral oil. It also serves as a physical barrier to protect samples from contamination during subsequent storage and processing. After PCR when the tubes cool the wax solidifies and samples can be taken by inserting a pipette tip through the wax layer. It may be possible to use an alternative source of paraffin wax such as that from Sigma-Aldrich which melts at 53–56°C.

Taq Bead™ hot-start polymerase

These small spherical beads supplied by Promega comprise wax encapsulating *Taq* polymerase that is released when the reaction reaches 60°C. Unlike the wax beads above the volume of wax is small and does not form a physical barrier above the reaction solution. The beads are suitable for use in either standard or heated-lid thermal cyclers, but for the former addition of a mineral oil overlay is necessary.

Magnesium wax beads

The presence of magnesium is essential for DNA polymerase activity. PCR reaction mixes set up in magnesium-free buffer can be activated at around 70°C when a StartaSphere™ wax bead (Stratagene) melts, releasing the correct amount of magnesium. These beads are small and nonbarrier-forming and so are compatible with heated-lid thermal cyclers. Mineral oil should be added for a non-heated lid thermocycler.

Booster PCR

The appropriate choice of annealing conditions allows primers to efficiently identify their complementary sequences when reasonable concentrations of DNA are being used, for example 1 µg of human genomic DNA (around 3×10^5 template molecules). However, at very low template concentrations, perhaps less than 100 molecules, the interactions between primers and template become less frequent. Instead there are more significant interactions between primers themselves, which can lead to primer-generated artifacts such as primer-dimers (Chapter 3). To enhance the specificity of template priming at low DNA concentrations a procedure called booster PCR can be employed (6). This involves performing the first few cycles of PCR at low primer concentration, so that the molar ratio of primer:template is around 10^7–10^8, the level normally found in a PCR (see *Table 2.2*). This enhances specific priming events and subsequently the concentration of primers can be 'boosted' during the amplification phase to maintain the 10^8 ratio of these reactants.

Cycle number and length

In general the number of cycles of PCR should be kept to the minimum required to generate sufficient product for further analysis or manipulation. This reduces the likelihood of errors arising and of nonspecific products accumulating. If the basic protocol (*Protocol 2.1*) does not yield sufficient product you could try to increase the amount of template in the first instance. Alternatively the number of cycles of PCR could be increased. It is possible to sample PCRs by removing an aliquot such as 0.1 vol (5 µl of a 50 µl vol) every 5 cycles at 25, 30 and 35 cycles during a 40-cycle reaction. The samples can then be analyzed by agarose gel electrophoresis to allow the appropriate number of cycles to be determined. This number will be the minimum number that gives good yield of a single product.

Another consideration that can influence PCR specificity is the time taken to move between temperatures during PCR cycling (7). Generally the faster the ramping rates the higher the specificity and the faster the reactions are completed. Some instruments can now achieve ramp rates of up to 2.5°C per s^{-1}.

Thermal cycler efficiency

It is easy to forget that instruments may malfunction. If your PCRs begin to fail then you should ask whether the thermal cycler is reaching the correct temperatures. Newer instruments have self-diagnosis features that can identify problems. As thermal cyclers get old they can tend to become less accurate in terms of their temperature profiles and often need adjustments. It is therefore worth using an independent temperature monitoring system occasionally, and certainly if variability in standard PCR results occurs. Temperature verification systems are available but are often very expensive. A simple and inexpensive temperature monitoring system can be made from a thermocouple, placed in a reaction tube containing water, and a digital thermometer. Even less expensive is to use a thin digital thermometer that fits inside a reaction tube filled with water. The temperature reached by the thermal cycler during PCR cycling can easily be monitored. In addition the system allows any temperature variation across the block to be assessed.

PCR optimization and additives

It is possible to purchase optimization kits that comprise a variety of buffers and additives to optimize conditions for PCR. For example, Stratagene produce an Opti-Prime™ PCR optimization kit comprising 12 different buffers and 6 additives, allowing a range of buffer conditions to be tested. Once optimized conditions have been determined the appropriate buffer can be purchased separately. Epigene also produce a Failsafe PCR optimization kit comprising a range of buffers.

Various 'enhancer' compounds have also been reported to improve the specificity or efficiency of PCR. These include chemicals that increase the effective annealing temperature of the reaction, DNA binding proteins and commercially available reagents. Such additives can be added to PCRs to enhance primer annealing specificity, reduce mismatch primer annealing and improve product yield and length. Additives that lead to a destabilization of base pairing can improve PCR particularly from difficult templates such as GC-rich sequences and may also increase specificity by their relatively greater destabilization of mismatched primer–template complexes. Although these compounds can be useful in some circumstances to improve suboptimal PCR conditions, some are not applicable to a wide range of templates and primer combinations. There is no 'magic' additive that will ensure success in every PCR and it may be necessary to test different additives under different conditions, such as annealing temperature. Such testing has been made easier with the advent of gradient thermal cyclers that allow the automatic testing of different annealing temperatures (Chapter 3). Compounds that have been added to PCR reactions include:

- dimethyl sulfoxide (DMSO), up to 10% (8);
- formamide at 5% (9);
- trimethylammonium chloride 10–100 µM (10);
- betaine (N,N,N-trimethylglycine) 1–1.3 M. A useful study and primary references are provided in Promega Notes http://www.promega.com/pnotes/65/6921_27/6921_27_core.pdf;
- nonionic detergents (11) such as Tween® 20 at 0.1–2.5%;
- polyethylene glycol 6000 (PEG) 5–15%;
- glycerol 10–15%;

- single-stranded DNA binding proteins such as Gene 32 protein (Amersham Pharmacia Biotech) added to 1 nM or *E. coli* single-stranded DNA binding protein at 5μM;
- 7 deaza-dGTP to reduce the strength of G–C base pairs; it is used at 150 μM with 50 μM dGTP as the G nucleotide mix;
- *Taq* Extender™ (Stratagene) increases *Taq* DNA polymerase DNA extension capacity leading to a greater proportion being fully extended. This is due to a reduction in the mismatch pausing when *Taq* DNA polymerase is dissociated from the template;
- Perfect Match® PCR Enhancer (Stratagene) apparently destabilizes mismatch primer template complexes where there are several mismatches close to the 3'-end. Perfect or near-perfect matched primer–template complexes including those with nonhomologous 5'-ends or tails are not destabilized and therefore generate good yields of product;
- Q-solution (Qiagen) modifies the melting behavior of template DNA, is used at a defined concentration for any template–primer combination and is not toxic.

The two additives that are probably most useful are DMSO, which disrupts base pairing and is usually added to 5–10% (v/v), and betaine (~1 M), which equalizes contributions of GC and AT base pairs towards duplex stability. It is advisable to adjust the denaturation, annealing and extensions temperatures down by perhaps 2°C when using betaine to adjust conditions for the weakening of the duplex bonding interactions and enzyme stability. In an interesting study undertaken by Promega they used NMR to analyse the constitutents of two commercially available PCR enhancer solutions and discovered that they were solutions of betaine (see http://www.promega.com/pnotes/65/6921_27/6921_27_core.pdf).

An example showing the effect of DMSO addition is shown in *Figure 4.3*.

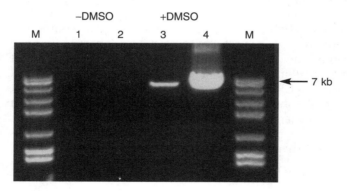

Figure 4.3

Agarose gel showing the effect of DMSO on PCR amplification. Lanes 1 and 2 show PCR amplification without DMSO, while lanes 3 and 4 show PCR amplification with 5% DMSO (final concentration). Lanes 1 and 3 are performed at 60°C annealing temperature and lanes 2 and 4 are performed at 58°C annealing temperature. (Provided by Dr Luis Lopez-Molina, Laboratory of Plant Molecular Biology, Rockefeller University.)

4.3 Template DNA preparation and inhibitors of PCR

PCR may be inhibited by a wide range of compounds derived from the biological specimens or method and reagents used to extract the DNA. Typical biological samples used for PCR are animal tissues and bodily fluids, including peripheral blood cells, urine, fecal samples, cell smears, hair roots, semen, cerebrospinal fluid, biopsy material, amniotic fluid, placenta and chorionic villus, bacteria, forensic and archaeological samples and plant tissues. Often PCRs are performed on relatively crude DNA preparations that contain unidentified inhibitory substances, and appropriate PCR controls are essential to eliminate the possibility of inhibition. Often if inhibition is occurring it is useful to dilute the DNA sample (*Figure 4.4*). This will have the effect of diluting both the template DNA and the inhibitor. If the inhibitor becomes diluted to a concentration that does not interfere with PCR then products should be obtained from the template DNA even if this requires a modest increase in the number of cycles. A common source of human DNA is blood, which should be collected into tubes containing 1 mg ml^{-1} EDTA to avoid coagulation. Heparin, a common anticoagulant, should be avoided, as it is a potent PCR inhibitor. Other substances in blood, perhaps porphyrin compounds, also inhibit PCR but can be removed by lyzing red blood cells and collecting the white cells by centrifugation for DNA preparation.

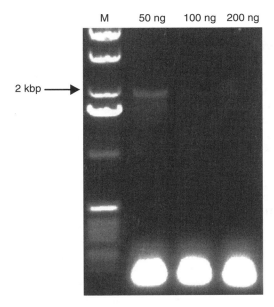

Figure 4.4

The effect of dilution of the DNA sample to enhance specificity of PCR. DNA concentrations are shown in ng. If the DNA sample contains a contaminant then diluting the sample can lead to dilution of the contaminant and successful amplification of the product. In this case the product can be seen only in the lowest dilution of the sample containing 50 ng DNA.

The variety of DNA sources demands a variety of extraction procedures that may include inhibitory compounds. For example, detergents are often used for cell lysis and protein denaturation. Nonionic detergents such as Triton X100, Tween 20 and Nonidet P40 usually do not inhibit PCR at concentrations up to 5%. By contrast, ionic detergents such as sodium dodecyl sulfate (SDS) are only tolerated at very low concentrations (< 0.01%) and should be removed by using a PCR clean-up kit or by phenol extraction and ethanol precipitation of the DNA before PCR. Adding 0.5% Tween 20 or Nonidet P40 may reverse inhibition at low SDS concentrations. Often extraction protocols use proteinase K to digest denatured proteins. It is important to remove or inactivate this protease before PCR, as *Taq* DNA polymerase is susceptible to digestion. Typically the sample is heated to 90–95°C and then phenol extracted and ethanol precipitated (see Chapter 6).

4.4　Nested PCR improves PCR sensitivity

Sometimes, no matter what you try, the specificity of your reaction may be very low, resulting in mispriming and the generation of false amplification products. Nested PCR provides a tool for increasing sensitivity allowing you to 'fish out' the specific amplification product from the 'sea' of nonspecific products. Since nested PCR is covered in more detail in Chapter 5 only the principle will be described briefly here.

Even if PCR primers have amplified nonspecific sequences, making it impossible to identify the desired product, it is highly unlikely that these nonspecific products will also have sites for a further pair of 'nested' specific primers. These nested primers are designed to anneal to sequences that will be present within the correct target PCR product. The true product will possess such nested primer target sequences while nontarget sequences will not. So, a second PCR using nested primers, designed to amplify an internal region of the original amplified product, should lead to a 10^4 enhancement of the true product over nonspecific products (*Figure 4.5*). Quite simply, a small aliquot of the first PCR, perhaps 1 µl of a 1-in-10 or 1-in-100 dilution can be used as the template for the nested PCR.

4.5　Contamination problems

Having considered how to optimize your reactions to reach the highest specificity and product yield, it is important to understand the potential problems of contamination. Great care should be taken to avoid contaminating your PCRs since this can be costly in terms of wasted time and reagents and potentially, in laboratories where clinical diagnosis or forensic analyses are performed, it could affect people's lives. The ability of PCR to amplify minute amounts of template DNA offers great advantages for detecting the target sequence, even in complex starting materials, but it has the disadvantage that small quantities of contaminating DNA may also be amplified. The analogy between PCR and good microbiological practice is a valid one. When dealing with potentially pathogenic bacteria, the scientist takes the necessary precautions to ensure that the bacteria are contained and do not contaminate the environment. In a similar way, the

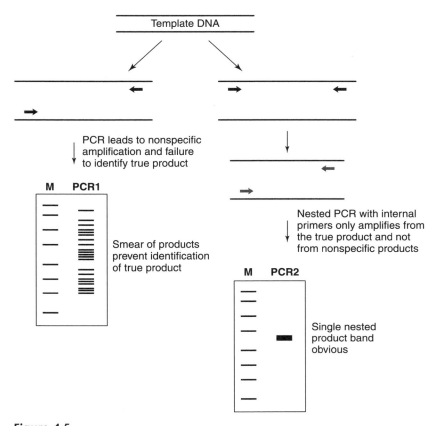

Figure 4.5

Schematic illustration of nested PCR from genomic DNA. The first-round PCR amplification results in a smear of nonspecific amplification products masking the target product. The second round of amplification uses a pair of nested primers that lie within the target region, leading to the amplification of only the correct product. M; molecular size markers. PCR1 and PCR2; amplified product(s) from the first PCR and the nested PCR respectively.

PCR is prone to contamination from DNA molecules present in the laboratory environment. There are some major sources of DNA contamination that can be introduced through bench surfaces, laboratory equipment, pipettors, airborne particles such as microbes and debris such as skin or hair, or contaminated solutions. The contamination may be due to new template such as microbial or particulate aerosols, often giving rise to a sporadic contamination of single samples.

More general contamination is due to:

- original template DNAs;
- cloned DNA molecules carrying the target gene;
- previously PCR-amplified molecules.

In general the same rules can be applied to prevent all types of contamination, although there are additional steps that you can take to prevent the latter form which is usually known as 'carryover' contamination. Also, the type of PCR experiments you are doing will affect the potential for contamination of your PCRs. For example, if you work in a research laboratory where different genes are being studied and manipulated, the problems of contamination are generally much easier to contain and, in the event of a problem, to track down and overcome. This contrasts with, for example, a clinical diagnosis laboratory, where multiple samples are routinely processed in the same way and hundreds of identical PCRs are performed to identify a particular genetic disease. Contamination in such an environment would be disastrous, and unfortunately is much more easily achieved, and so requires the most rigorous steps to prevent any possible problems.

Probably the most serious source of contamination is carryover from a previously processed PCR. However, there are correct ways of setting up a PCR and relatively simple procedures for ensuring that any contaminating molecules are destroyed. You should set yourself a high standard and should aim to become a perfect practitioner of the PCR. Do not be put off by the person on the next bench who 'just throws everything in' or even worse 'never bothers about controls'. They may try to persuade you to adopt their slap-dash methods because it will save you time. Do not be tempted. What you are doing now may not be too sensitive to contamination but you may be doing something very different next month or next year, and good practice that becomes second nature now will prevent serious errors later. A short article by Kwok and Higuchi (12) describes probable contamination problems and gives some useful advice on prevention. To highlight the potential problem they calculated that if you were to dilute a completed 100 µl PCR reaction into the water of an Olympic-sized swimming pool and then, after mixing, were to remove a fresh 100 µl aliquot, it would contain some 400 molecules of DNA.

Control reactions

You should routinely set up control reactions as described in Section 4.1. In addition it is a good idea to set up the control reactions as the last in a series so that any cross-contamination can be detected; for example if contamination of a premix (see 'Premixes', below) occurs during the PCR set-up then the early reactions may not be contaminated, but later ones may be. This will be detected if controls are set up last.

A designated PCR set-up area

It is best to try to avoid setting up PCRs in the same area as you process post-PCR samples. Small PCR cabinets or chambers with built-in UV lamps to destroy template DNA and maintain a contamination-free environment are available. Alternatively, set up your PCRs in another room, preferably one dedicated for that purpose. If these options are not available then use a separate bench in the laboratory as a PCR area. Whatever solution you come up with for keeping the set-up and post-PCR areas separate, follow the rule that 'separation helps prevents contamination'. Always wear fresh gloves when setting up reactions and work on clean surfaces regularly washed down with 1% sodium hypochlorite and ethanol.

Pipettes and tips

Even moving to another building will not prevent contamination problems if it is you that takes the contaminants with you! Likely contamination sources are aerosols from opening tubes containing DNA or from pipettes. The latter will be a particular problem if you only have one set of pipettes that you use for both setting up PCRs and for the manipulation of the resulting solutions. It is therefore important to have a dedicated set of pipettes for setting up the reactions. Additional measures to ensure clean pipetting will, to a large extent, be governed by your budget. A simple way of trying to keep things clean is to regularly wipe the tip of the pipette barrels with ethanol-soaked tissue. However, do not soak the pipette in ethanol or suck any up into the barrel, because it will be added to the reaction next time you perform a pipetting step. Various special tips are available that contain a filter to prevent any exchange of aerosol or solution between the pipette barrel and the solution being pipetted. Such tips work well, although they are more expensive than ordinary pipette tips. Even if you cannot afford to use them for every pipetting step there are several stages of setting up PCRs where they are invaluable:

- for removing aliquots of oligonucleotides from stock solutions, since the contamination of a primer stock is a very serious issue resulting in the need to resynthesize the oligonucleotide with the inevitable time delay and expense that entails;
- template DNA additions to prevent contaminating your DNA stock.

Finally, positive displacement pipettes are available that employ a disposable piston that makes direct contact with the solution being aliquoted so there is no air/liquid interface that might allow contamination in a normal air displacement pipette. They also allow very accurate delivery of the small volumes. These are relatively expensive, require special tips but are a useful investment.

Solutions

Most enzyme suppliers now provide appropriate buffer stocks, magnesium solutions and even dNTP mixes with their enzymes. It can be helpful to aliquot these reagents on arrival to ensure protection from contamination during routine use. For other solutions, always use the highest quality of reagents available and make up solutions using autoclaved ultrapure water. Obviously if any solution is contaminated with DNA then it will contaminate the reaction(s) to which it is added. As with any DNA work, solutions should be autoclaved where possible and, again, aliquoted into appropriate volumes (often 0.1–0.2 ml) for storage purposes. This allows a single tube to be taken from the freezer for PCR set-up while the rest of the solution remains safely frozen. After use the tube can be discarded and a fresh aliquot used for the next set of PCRs. If you encounter any contamination problems, throw away any working stocks and use fresh aliquots from your stocks.

Premixes

Where possible you should prepare a premix containing the common reactants for all the PCRs you are setting up. This reduces the number of repetitive pipetting steps and thus potential cross-contamination. Calculate the required amounts of each reaction component, multiply by the number of reactions to be set up and add one or two additional reaction volumes to account for minor pipetting errors. For example, if you are setting up 10 reactions then premix sufficient reactants for 11 or 12 reactions. There is a useful online calculator to help prepare premixes at http://www.sigmaaldrich.com/ Area_of_Interest/Life_Science/ Molecular_Biology/PCR/Key_Resources/PCR_ Tools.html; remember to add an extra 1–2 reactions when doing the calculation. Obviously, when preparing premixes, you should take care to prevent contamination while transferring aliquots from stock solutions. Aliquot the premix into reaction tubes prior to the addition of other reactants. Remember to set up the controls last.

Mineral oil

The use of oils is now less common since most thermal cyclers have hot-lid technology (Chapter 3). If you still use mineral oil it is important to ensure it does not become contaminated by careless use of a pipette. Pour mineral oil from the stock bottle into a microcentrifuge tube for dispensing and then discard this aliquot. An alternative is silicone oil,which creates a flatter and clearer phase transition interface. It is available in a dropper bottle format from Life Technologies/Invitrogen.

Other sources of contamination

There can be particular problems associated with human DNA analysis. A microscopic flake of skin or even dandruff is sufficient to contaminate a PCR, potentially leading to an erroneous result. Always wear gloves when setting up PCRs and ideally use an appropriate PCR hood.

4.6   Preventing contamination

Uracil N-glycosylase

Substituting dUTP for dTTP during PCR provides one route for eliminating carryover of previous PCR products. Such dU-containing PCR products are susceptible to the enzyme uracil N-glycosylase (UNG), which will hydrolyze uracil glycosidic bonds in both single- and double-stranded DNA. This releases uracil, resulting in strand cleavage when the DNA is heated during PCR. The enzyme does not affect RNA or normal DNA molecules. It is therefore an easy task to destroy previous PCR products by including a uracil N-glycosylase pretreatment of the reactant mix prior to PCR. Such a practice is particularly critical in diagnostic laboratories where routine amplification of identical products is being performed on a daily basis and the possibility of cross-contamination is omnipresent, but must be avoided to prevent false positives and misdiagnosis.

The resulting PCR products contain dU in place of dT and will behave normally for most forms of post-PCR analysis. For example such products are as effective as dT-containing templates in hybridization protocols and DNA sequencing. In addition the products can be cloned efficiently, although they must be transformed into an *ung⁻* strain of *E. coli* lacking the normal uracil *N*-glycosylase activity that would otherwise destroy the cloned DNA. Restriction enzyme digestion is efficient for common enzymes such as *Eco*RI, and is only slightly reduced for others such as *Hin*dIII, although for many enzymes the efficiency of digestion remains to be determined. Major applications for which dU-containing templates are not recommended include those involving molecular interactions studies, such as DNA binding assays.

Some proofreading DNA polymerases are inefficient at incorporating dUTP into PCR products and therefore this approach cannot be used with such enzymes, although it is very effective with *Taq* DNA polymerase. The enzyme can retain some activity even after treatment at 95°C and therefore some consideration of this may be necessary if dUTP incorporation is being used in a PCR previously treated with UNG. It is recommended that the reaction be held at 72°C, or be frozen at –20°C if it is not to be analyzed immediately. Several manufacturers supply uracil *N*-glycosylase specifically for this purpose, including AmpErase® (Roche) and HK-UNG™ Thermolabile Uracil N-Glycosylase (EpiCentre).

It has also been reported that incorporating dUTP in place of dTTP during primer synthesis can prevent carryover contamination. Treatment of the PCR products with UNG will destroy the original primer regions of the PCR product thereby preventing amplification of the full-length product during a subsequent PCR (13).

UV irradiation

UV treatment of reaction components by illumination at 254 nm has been used to overcome contamination. This treatment leads principally to thymidine dimerization resulting in the inability of the DNA to function as a template during the PCR. Remember that the reaction components must be treated *before* addition of the template DNA or thermostable DNA polymerase! UV irradiation systems, which are often sold predominantly for their ability to UV crosslink DNA to nylon membranes, can also be used to pretreat PCR components. Companies supplying such crosslinkers include Stratagene, GE Healthcare, Hoefer Inc., Ultra-Lum, UVP and Cole-Palmer. UV irradiation does not completely destroy contaminating DNA but it dramatically reduces the concentration, often by a factor of 10^4. UV has little effect upon the dNTPs and, perhaps surprisingly, does not appear to affect the short primers significantly. However, this also means that it will be less efficient at inactivating short contaminating DNAs (<250 bp) than longer DNAs. The UV approach has had a mixed reception and it is not clear how widely used it is.

Psoralen and isopsoralen treatment

An alternative approach to prevent contamination carryover by pretreatment of PCR reactants has been developed (14). Psoralens such as

8-methoxypsoralen (8-MOP) intercalate into double-stranded nucleic acids and result in interstrand crosslinks upon longwave UV irradiation (300–400 nm). Treatment of the reaction components (but not the template DNA) with 8-MOP and UV irradiation leads to contaminating DNA molecules becoming crosslinked and unable to denature during PCR and therefore these molecules cannot act as templates. The treatment does not adversely affect the primers or DNA polymerase.

A post-PCR treatment can also be used to prevent subsequent carryover of PCR products. Isopsoralens, for example 4′-aminomethyl-4,5′-dimethyl-isopsoralen (4′-AMDMIP) or 6-aminomethyl-4,5′-dimethylisopsoralen (6-AMDMIP), form covalent monoadducts with double-stranded DNA when photo-activated by longwave UV irradiation (15). DNA polymerases are unable to proceed when they reach such a modified base and so contaminating DNA cannot act as template in a subsequent PCR. The treatment involves addition of the isopsoralin at the start of the PCR and a post-PCR irradiation to modify the products to ensure they are unable to contaminate further PCRs.

4.7 Troubleshooting guide

This Section is designed to help identify possible causes for common PCR problems and to suggest solutions that may overcome the problem. Generally if PCR does not work then there is likely to be something wrong with the template DNA, primers, polymerase or the choice of conditions. Initially, it is worth trying it again under the same conditions to ensure that there was not a simple error that resulted in the failure. In addition it is important to include controls. A positive control with a template known to amplify well will ensure that all reagents are added and that they are all functioning. Negative controls, one lacking template DNA and another lacking primers, will reveal any contamination, disclose whether primer-dimer formation is an issue and ensure no nonspecific priming events are occurring. It can also be worth starting from the beginning again by scrapping all the reagents you have used and preparing fresh ones. *Table 4.1* provides some guidelines for helping solve problems that you may face with your PCRs.

Table 4.1   PCR troubleshooting guide

Problem	Possible reasons	Things to try
No PCR product	One or more component missing or faulty	Use a checklist to ensure all components are added to tubes. Check the concentrations of *all* reagents, including primers, template, dNTPs, $MgCl_2$ and buffer. If possible use a premix containing most of the reagents to reduce pipetting steps. Always perform a positive control to show that reagents are functioning efficiently
No PCR product	Insufficient number of cycles	Increase the number of cycles. Reactions can be sampled by removing an aliquot (0.1 vol i.e. 5 µl of a 50 µl vol) every 5 cycles (at 25, 30 and 35 cycles) during a 40-cycle reaction to allow the appropriate number to be determined

Table 4.1 *continued*

Problem	Possible reasons	Things to try
No PCR product	Poor primer design	Reconsider the design of primers and check DNA sequencing data. If possible test primer function in a DNA sequencing reaction to ensure it is uniquely priming from the desired position in the template DNA. Resynthesize longer primers with particular care to ensure 3'-end is perfectly matched to template
No PCR product	Template quality or quantity	Too little template or template contaminated with inhibitors or of poor quality through, for example, long-term storage in dilute solution may all lead to lack of product. Check the integrity of the template by agarose gel electrophoresis. Dilute a fresh aliquot of template DNA and set up a series of concentrations of template including higher and lower concentration, the former in case there was too little and the latter to dilute out any inhibitory substances. Alternatively prepare a fresh sample of template DNA
No PCR product	Thermal cycler malfunction	Check that the thermal cycler is reaching the preset temperatures by using an inexpensive thermocouple device to monitor reaction tube temperature, and that it is cycling correctly
No PCR product	Insufficient denaturation or difficult templates	An initial incubation at 95°C for 5 min should be sufficient to denature the template. For highly GC-rich sequences it may be necessary to add components that enhance denaturation and destabilize the DNA duplex. See Section 4.2: Temperatures
No PCR product	Extension time too short	As a rule use 1 minute for each kbp of target. Increase the time in 1 min increments if no products are observed
No PCR product	Enzyme problem	This should become clear from your positive control. 1–2 units of enzyme is usually sufficient for most applications. If the enzyme is old and has been poorly stored then it is likely to be suboptimal. Test another aliquot or batch of enzyme. In rare cases adding more units may yield a product, although it may also increase background
No PCR product	Buffer components	If you use a 10 × PCR buffer supplied by the enzyme manufacturer this should not be the problem. If possible test another batch of buffer. The magnesium ions may affect product yield if added at the wrong concentration. Remember that many manufacturers supply PCR buffers that are magnesium-free so check that you are adding the correct range of magnesium to give a final concentration of typically 1.5 mM but within the range 1–5 mM normally
No PCR product	Deoxynucleotides	Ultrapure solutions should perform well and the dNTPs are present in the reaction in vast excess, therefore even if they are added at a lower concentration some product would be expected. Check the concentration you are adding and ensure the positive control gives a product. Test a new aliquot or batch of dNTPs

Table 4.1 *continued*

Problem	Possible reasons	Things to try
Multiple products generated	Low annealing temperature	Use a hot-start procedure (Section 4.2) and increase the annealing temperature in 2°C steps. The use of a gradient thermal cycler is recommended. Alternatively use a touchdown procedure (Section 4.2)
Multiple products generated	Too many cycles	Reduce the number of cycles used. Reactions can be sampled by removing an aliquot (0.1 vol i.e. 5 μl of a 50 μl vol) every 5 cycles (at 25, 30 and 35 cycles) during a 40-cycle reaction to allow the optimal number to be determined
Multiple products generated	Poor primer design	Reconsider the design of primers and check DNA sequence data. If possible test primer function in a cycle or manual DNA sequencing reaction. Resynthesize longer primers with particular care to ensure that the 3'-end is perfectly matched to the template
Smeared products generated	Extension time too long	Reduce the extension time in 30 s to 1 min steps
Smeared products generated	Too many cycles	Reduce the cycle number in five-cycle steps or use the sampling technique described above
Smeared products generated	Denaturation temperature too low	Check that the thermal cycler reaches the denaturation temperature using a thermocouple and if necessary increase the denaturation temperature in 1°C steps
Smeared products generated	Too much polymerase	Reduce the amount used
Smeared products generated	Too much template	Reduce the amount of template added, ideally by testing a dilution series
Faint product band	Insufficient number of cycles	Increase the number of cycles. Reactions can be sampled by removing an aliquot (0.1 vol, i.e. 5 μl of a 50 μl vol) every 5 cycles (at 25, 30 and 35 cycles) during a 40-cycle reaction to allow the appropriate number to be determined
Faint product band	Template quality or amount	Too little template or template contaminated with inhibitors. Dilute a fresh aliquot of template DNA and set up a series of concentrations of template including higher and lower concentrations, the former in case there was too little and the latter to dilute out any inhibitory substances
Faint product band	Extension time too short	As a rule use 1 min for each kbp of target. Increase the time in 1 min increments if no products are observed
Faint product band	Additive required	Add a PCR enhancer compound (see Section 4.2)

Further reading

Boleda MD, Briones P, Farreo J, Tyfield L, Pi R (1996) Experimental design: a useful tool for PCR optimization. *BioTechniques* 21: 134–140.

Dieffenbach CW, Dveksler GS (1993) Setting up a PCR laboratory. *PCR Methods Appl* 3: S2–7.

Dragon E (1993) Handling reagents in the PCR laboratory. *PCR Methods Appl* **3**: S8–9.
Roux KH (1995) Optimization and troubleshooting in PCR. *PCR Methods Appl* **4**: S185–194.

References

1. Blanchard MM, Taillon-Miller P, Nowotny P, Nowotny V (1993) PCR buffer optimization with uniform temperature regimen to facilitate automation. *PCR Methods Appl* **2**: 234–240.
2. Don RH, Cox PT, Wainwright BJ, Baker K, Mattick JS (1991) 'Touchdown' PCR to circumvent spurious priming during gene amplification. *Nucleic Acids Res* **19**: 4008.
3. D'Aquila RT, Bechtel LJ, Videler JA, Eron JJ, Gorczyca P, Kaplan JC (1991) Maximizing sensitivity and specificity of PCR by pre-amplification heating. *Nucleic Acids Res* **19**: 3749.
4. Erlich HA, Gelfand D, Sninsky JJ (1991) Recent advances in the polymerase chain reaction. *Science* **252**: 1643–1651.
5. Ruano G, Pagliaro EM, Schwartz TR, Lamy K, Messina D, Gaensslen RE, Lee HC (1992) Heat-soaked PCR: an efficient method for DNA amplification with applications to forensic analysis. *BioTechniques* **13**: 266–274.
6. Ruano G, Fenton W, Kidd KK (1989) Biphasic amplification of very dilute DNA samples via 'booster' PCR. *Nucleic Acids Res* **17**: 5407.
7. Wittwer CT, Garling DJ (1991) Rapid cycle DNA amplification: time and temperature optimization. *BioTechniques* **10**: 76–83.
8. Seto D (1990) An improved method for sequencing double stranded plasmid DNA from minipreps using DMSO and modified template preparation. *Nucleic Acids Res* **18**: 5905–5906.
9. Sarkar G, Kapelner S, Sommer SS (1990) Formamide can dramatically improve the specificity of PCR. *Nucleic Acids Res* **18**: 7465.
10. Hung T, Mak K, Fong KA (1990) Specificity enhancer for polymerase chain reaction. *Nucleic Acids Res* **18**: 4953.
11. Bachmann B, Luke W, Hunsmann G (1990) Improvement of PCR amplified DNA sequencing with the aid of detergents. *Nucleic Acids Res* **18**: 1309.
12. Kwok S, Higuchi R (1989) Avoiding false positives with PCR. *Nature* **339**: 237–238.
13. Longo MC, Berninger MS, Hartley JL (1990) Use of uracil DNA glycosylase to control carry-over contamination in polymerase chain reactions. *Gene* **93**: 125–128.
14. Jinno Y, Yoshiura K, Niikawa N (1990) Use of psoralen as extinguisher of contaminated DNA in PCR. *Nucleic Acids Res* **18**: 6739.
15. Cimino GD, Metchette KC, Tessman JW, Hearst JE, Isaacs ST (1991) Post-PCR sterilization: a method to control carryover contamination for the polymerase chain reaction. *Nucleic Acids Res* **19**: 99–107.

Analysis, sequencing and *in vitro* expression of PCR products

5

5.1 Introduction

Analysis of PCR products is critical in optimizing PCR conditions to yield reliable and accurate results, and in interpreting the levels of products generated, for example, in a diagnostic context.

The first part of this Chapter considers how to analyze PCR-amplified DNA fragments in the most appropriate way for different experimental strategies. In general it is important to ensure that you develop robust and reproducible PCR conditions, particularly if they are to become part of a routine high-throughput screening procedure. Approaches for detection and analysis of PCR products and for verification of product identity, including direct sequencing strategies, are considered. Following this, procedures for quantitation of a specific product are covered. The final section deals with *in vitro* expression of PCR products to yield protein products. Real-time analysis is covered in Chapter 9.

5.2 Analysis of PCR products

There are many ways to analyze PCR products depending upon the information required:

- the presence or absence of the target DNA sequence;
- the length of the amplified fragment;
- the yield of PCR product to quantitate the relative or absolute amounts of the starting DNA or RNA;
- sequence analysis, either by differential probe hybridization or by direct sequencing of the product.

It will often be possible to predict the size(s) of expected PCR products, which are of a length defined by the positions of the PCR primers. In other cases the size of product cannot be predicted, for example in some cDNA amplification experiments such as RACE-PCR and in some genomic cloning or walking experiments. Often PCR products are relatively small, in the range of 0.2–3 kbp, and even in many genomic cloning experiments they are less than 10 kbp in length. The simplest and most direct methods to analyze PCR products involve gel electrophoresis.

Gel electrophoresis

The most common and rapid way of analyzing PCR products is by standard agarose gel electrophoresis. Depending on the expected size of the amplified fragment, a fraction of your PCR reaction should be loaded onto a 0.8–3% agarose gel containing 1 μg ml⁻¹ ethidium bromide. Usually one-tenth or one-fifth of the reaction volume is loaded and the remainder is stored at 4°C or –20°C for subsequent use. An aliquot of loading dye containing glycerol and a marker such as bromophenol blue should be added to the sample to assist both loading on the gel and visualization of the sample migration through the gel. If you have used a reaction mix already containing a dye, such as described in Chaper 3, then the sample can be loaded directly.

Appropriate molecular size markers such as a 100 bp or 1 kbp DNA ladder or bacteriophage lambda or φX174 restriction enzyme digests, available from a range of manufacturers, or previously characterized PCR products, should be loaded in adjacent lanes of the gel. The amplified fragment(s) should be readily visible under ultraviolet transillumination (always use protective eye-wear and minimize time of exposure) and the gel can be photographed using a camera or digital imaging system to record the results. In most cases a 1% agarose gel gives sufficient resolution for DNA fragments between 500 and 4000 bp. If you are expecting very small fragments then it is probably better to use a specialized agarose such as Metaphor® at 3.5% which has very high resolution (10–1000 bp), or NuSeive® GTG at 4% (50–2000 bp), both from Cambrex Bioproducts. The latter allows efficient DNA recovery. Other specialty agarose preparations are also available such as Agarose 1000 (Invitrogen), which provides resolution of up to 10 bp for PCR products up to 350 bp in length when used at a concentration of 4.5%. Such gels can be useful for analysis of multiplex PCRs that contain several PCR products. For very small products, or for identifying small size differences between products, such as in microsatellite repeats or single-strand conformational polymorphisms, nondenaturing or denaturing polyacrylamide gels provide the most appropriate resolution system (Chapter 11).

If your PCR conditions are optimal and your PCR has worked well you should be able to visualize an intense sharp band of the expected size. Sometimes you may also observe small primer-dimer products at the leading end of the gel (Chapter 3). Frequently these products are most pronounced in the absence of a specific PCR product. If the PCR conditions were not optimal or the reaction used degenerate primers (Chapter 3) and a complex source of template DNA (such as human genomic DNA) you may see additional bands that are most probably due to nonspecific priming events. Common reasons for the occurrence of such products include low annealing temperatures, high Mg²⁺ concentrations and/or the occurrence of similar priming sequences in the complex template source. Usually it is fairly straightforward to determine which 'band' contains the correct DNA fragment based on its expected size, if known, and its sharper appearance and higher intensity compared to the lower intensity of nonspecific amplification products. However, sometimes it is difficult to distinguish between nonspecific and specific amplification products either due to similar band intensities or due to the presence of a smear of DNA amplification products. Smearing of DNA amplification products is most often associated with nonspecific primer annealing conditions, poor quality DNA or low copy number template, or a combination of such factors (Chapter 4). In such cases it is often possible to increase the sensitivity of the analysis in order to identify the amplified target DNA by, for example, nested PCR or Southern or dot/slot hybridization (Section 5.3). Such methods can assist in the optimization of PCR conditions so that you are able to amplify the desired product routinely and reproducibly, allowing the use of homogenous detection methods that do not rely on gel fractionation.

5.3 Verification of initial amplification product

Often a PCR product will be used for subsequent experiments and so it is important to ensure that the amplified DNA fragment really represents the DNA sequence of interest. This Section covers hybridization analysis, nested PCR and restriction analysis, which are all approaches to verify product identity that can be more rapid for processing a number of samples than the most direct approach; direct DNA sequence analysis of the PCR product (Section 5.4)

Southern and dot blot analysis

Southern blot analysis involves the transfer of DNA fragments from an agarose gel to a nylon membrane by capillary transfer, followed by DNA hybridization with a specific probe to detect the presence of the target DNA fragment (1). It offers a sensitive approach for the detection of the target sequence using probes that are either radiolabeled or nonisotopically labeled, including enzyme-linked detection systems. DNA hybridization conditions can be controlled at both the hybridization and post-hybridization stages by altering the temperature and salt concentration. The use of a probe is more sensitive than ethidium bromide detection methods and can reveal a target fragment that was not visible on the original ethidium bromide-stained gel. In addition when the probe hybridizes it confirms the identity of the fragment. Although homologous probes from the target gene are preferred, heterologous probes obtained from a similar gene from another organism also work well, but may require more optimization and less stringent hybridization and post-hybridization conditions.

An alternative and more rapid technique than Southern blot analysis is dot or slot blotting. Here a sample of the amplification reaction is transferred directly to a membrane followed by DNA hybridization to a specific probe. It does not involve agarose gel electrophoresis or capillary transfer and so is more rapid. Although dot or slot blotting identifies the presence of the correct amplification product this technique does not determine its size or the presence of other PCR products. Dot or slot blot analysis is often used when there are large numbers of PCR samples to be analyzed.

For Southern and dot blot hybridization a range of probe-labeling strategies are available. Oligonucleotide probes are generally 5′-end labeled with ^{32}P by T4 polynucleotide kinase-catalyzed transfer of terminal labeled phosphate from [γ–^{32}P]ATP to the 5′-end of the oligonucleotide. Larger DNA fragments are often labeled by nick translation or random hexamer-primed labeling with the incorporation of ^{32}P from [α^{32}P] dCTP or dATP during DNA synthesis by a suitable DNA polymerase such as T7 DNA polymerase. Probes may also be labeled nonisotopically with a range of fluorescent dyes, with crosslinked enzymes such as horseradish peroxidase (HRP) or alkaline phosphatase (AP), with digoxygenin (DIG), which is detected by a specific anti-DIG antibody coupled to HRP or AP, with acridinium esters or with other tags.

Nested PCR

Nested PCR offers a quick and reliable way of verifying a PCR product. It generally uses two primers that are internal to the product of the first PCR. The PCR product from the first PCR is used as template DNA for a second round of PCR with the internal primers. This should yield a smaller PCR product compared with the original product (*Figure 5.1*). It is estimated that nested PCR leads to a 10^4-fold increase in sensitivity of detection of the correct product. Even if the first round PCR product is poorly represented

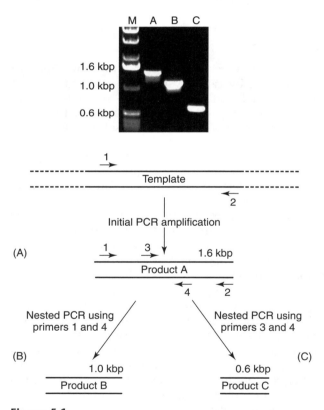

Figure 5.1

Diagrammatic representation of nested PCR and analysis by agarose gel electrophoresis. (A) Initial PCR amplification product using two original flanking primers (primers 1 and 2); (B) nested PCR from the primary PCR amplification product using one original flanking primer (primer 1) and one internal nested primer (primer 4); (C) nested PCR from the primary PCR amplification product using two internal nested primers (primers 3 and 4). M denotes molecular size markers.

within a background of nonspecific products it will be enriched for the specific template allowing efficient amplification by the nested PCR primers. By contrast the nonspecific products of the first PCR are unlikely to have sequences that are complementary to the nested primers and so there should be no nonspecific amplification after the nested PCR.

Even if it is not possible to design two internal primers because of lack of sequence information, for example when only limited peptide sequence data are available, it is usually still possible to perform a nested PCR. One new internal primer could be used together with one of the original primers. Alternatively, extending one or both of the original primers by even two or three nucleotides at their 3'-ends should be sufficient to impose increased specificity on the nested PCR. As discussed in Chapter 3 it is the 3'-end of the PCR primer that is most critical for determining specificity of PCR amplification. If the 3'-nucleotide is not complementary to the template then no amplification should occur. So extending a nested PCR primer by two or three nucleotides should allow the specific target to be amplified but not the nonspecific products even though the nested primers overlap significantly with the original primers. Of course in this case the use of an enzyme with 3'-exonuclease proofreading activity should be avoided to protect the differentiating 3'-end. An example of the design of original and nested PCR primers by back-translation of a limited region of amino acid sequence information is shown in *Figure 5.2*.

To reduce manipulations and avoid any contamination problems both the initial and nested PCR reactions can be performed in a single tube. Both primer pairs are included at the start of the PCR but the nested primers are designed to have a lower T_m than the initial primer pair. This allows amplification of the primary target at an annealing temperature above that of the nested primers. Then, a second PCR program is performed but at a lower annealing temperature, allowing the nested primer pair to amplify the specific PCR product from the initial PCR product. The PCR products can then be analyzed by agarose gel electrophoresis and should reveal both the primary amplification product and the smaller nested amplification product. However, if the primary amplification resulted in multiple bands or a smear the nested amplification product may be harder to identify. It is best not to use the initial PCR product for further experiments since the extended number of PCR cycles increases the chances of PCR-generated mutations.

An obvious potential problem when verifying the identity of the initial PCR products by nested PCR is the presence of the original template DNA. If the initial product is nonspecific, but sufficient original template is present to allow amplification by the nested primers, a positive result may lead to the erroneous assumption that the initial PCR product represents the correct target product. To avoid any amplification from the original template the first PCR can be diluted so that the absolute amount of original template is negligible. In a case where there is a defined initial PCR product then a more reliable approach is to physically purify the PCR product from the original template DNA, for example by agarose gel electrophoresis and gel purification (Chapter 6).

In any PCR experiment it is important to perform suitable controls to ensure specificity of the PCR. In nested PCR the increased sensitivity of the

```
Amino acid sequence        A   D   T   E   W   D   K   G   E   H   G
DNA sequence          NNNGCAGACACAGAATGGGACCAAGGAGAACACGGANNNN
                           G   T   G   G       T   G   G   G   T   G
                           C       C                   C           C
                           T       T                   T           T


Primer for PCR 1           GCAGACACAGAATGGGACAAAGG
(256)                 5'    G   T   G   G       T   G   3'
                           C       C
                           T       T


Primer for nested PCR              GAATGGGACAAAGGAGAACACG
(128)                        5'    G       T   G   G   G   T  3'
                                                   C
                                                   T
```

Figure 5.2

Design of degenerate primers from amino acid sequence data. The primer mix for initial PCR represents a combination of 256 different sequences and is used together with an appropriate downstream primer in PCR 1. Due to the limited amount of amino acid sequence data available the nested primer (128 different sequences) overlaps with part of the PCR 1 primer, but has been extended so that the 3'-end is different.

method makes this much more critical as any contamination will be enhanced. It is essential to include single primer control reactions to ensure primer specificity as described in Chapter 4, as well as no DNA and no primer controls.

Restriction analysis of a PCR product

Restriction digest analysis of PCR products is not commonly used to verify identity. However, the approach can be efficient giving a clear result and is relatively rapid and simple requiring mixing of an aliquot of PCR product, 10× restriction buffer and restriction enzyme, incubation to allow digestion and then agarose gel electrophoresis to visualize the restriction fragments. Of course it is only useful if a restriction map of the amplified DNA fragment is available. Not all restriction enzymes are active in the presence of various PCR components so an additional purification step may be required. Direct restriction analysis can be useful for verifying site-specific mutations that introduce or remove a restriction site from a PCR product. The approach can be coupled with Southern blot hybridization methods for product identification and can be used to analyze nested PCR products.

In summary, nested PCR offers a rapid and sensitive approach for verifying PCR amplification profiles. However, Southern blot data obtained under high stringency conditions offer more definitive verification of product identity. Some combination of approaches may be required in difficult cases. Of course the most definitive confirmation of identity of a PCR product is determination of its DNA sequence, a process that can be more rapid than Southern blot analysis if small numbers of samples are involved (Section 5.4).

5.4 Direct DNA sequencing of PCR products

Once a PCR product has been cloned into a suitable vector the recombinant molecule can be used for DNA sequence analysis of the PCR insert. However, during product verification, particularly where there are multiple samples to screen, it is not always efficient to clone each fragment. A more direct approach is to perform direct sequence analysis of the PCR product (2,3). It might be argued that this approach should be routinely used as the only method of PCR product verification, however, it is not always straight-forward and can involve greater time and effort than less direct methods such as nested PCR, in particular when processing large numbers of samples. Nonetheless, with improvements in automation and sequencing tech-nologies (see below) and the ever-decreasing cost it seems reasonable to assume that sequencing will eventually become the preferred approach to product verification.

It is also important to remember that direct sequencing provides an additional benefit in that you are sequencing a population of PCR molecules. Since errors can occur randomly during PCR any single clone is derived from only one PCR product that may or may not represent the true natural sequence. It is therefore usually necessary to sequence several independent clones to ensure a correct consensus sequence is obtained. In direct sequencing one is determining such a consensus sequence directly. Only if the PCR is performed on a very small amount of template is there likely to be a risk that an early PCR error will be detected in the final product population. However, the reproducibility of direct DNA sequence data will also depend upon the source of template DNA. In most cases of DNA isolation from fresh samples there will be no difficulties, but for old samples in which the DNA may be damaged, more care may be required. A study of old forensic samples indicated that the level of errors was 30-fold higher than in control samples, effectively leading to an error rate as high as 1 in 20 nucleotides (4). It was demonstrated by HPLC and ionization mass spectrometry that there was a decrease in the concentrations of the four normal bases, and an increase in oxidation products within the old DNA samples. It was found that both strands of DNA should be sequenced, and replicate PCRs should be performed and sequenced from the same DNA samples. Similar arguments would apply to other aged samples, such as those used for PCR archaeology (Chapter 3).

DNA sequencing chemistry and automation

Dideoxy terminator DNA sequencing (5) involves the incorporation of 2′,3′-dideoxynucleotide 'terminators' into nascent DNA chains (5; *Figure 5.3*). Basically, a DNA sequencing reaction results in DNA polymerase-directed synthesis of new DNA from a primer annealed to a single-strand template DNA molecule. In general for PCR products the template will be double-stranded and is heat denatured and rapidly frozen by placing in dry ice or liquid nitrogen to prevent reannealing of the separated strands. Various DNA polymerases can be used for sequencing reactions including T7 DNA polymerase (6) and thermostable enzymes such as *Taq* (7) or AmpliTaq™ (PE Biosystems), Vent$_R$® exo⁻ (New England Biolabs), *Pfu* exo⁻ (Stratagene)

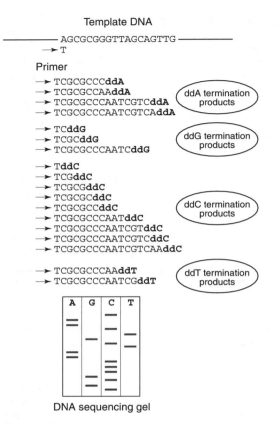

Template DNA

AGCGCGGGTTAGCAGTTG

→ T

Primer

→ TCGCGCCCddA
→ TCGCGCCAAddA   ddA termination
→ TCGCGCCCAATCGTCddA   products
→ TCGCGCCCAATCGTCAddA

→ TCddG
→ TCGCddG   ddG termination
→ TCGCGCCCAATCddG   products

→ TddC
→ TCGddC
→ TCGCGddC
→ TCGCGCddC
→ TCGCGCCddC   ddC termination
→ TCGCGCCCAATddC   products
→ TCGCGCCCAATCGTddC
→ TCGCGCCCAATCGTCddC
→ TCGCGCCCAATCGTCAAddC

→ TCGCGCCCAAddT   ddT termination
→ TCGCGCCCAATCGddT   products

DNA sequencing gel

Figure 5.3

Dideoxynucleotide chain termination approach for DNA sequencing. An oligonucleotide primer is extended by DNA polymerase that incorporates the appropriate dNTPs. Occasionally the polymerase incorporates a dideoxy NTP that lacks a 3′-OH group and therefore cannot support further nucleotide addition. This chain is therefore terminated by the ddNTP. Within the population of molecules will be chains terminated at each position. A high-resolution system based on polyacrylamide gel or capillary electrophoresis separates the products and allows the DNA sequence to be read. Either primers or ddNTPs can be fluorescently labeled allowing detection in an automated DNA sequencer.

and *Bst* DNA polymerase I (8) (BioRad). The DNA polymerase uses the four dNTPs (dATP, dCTP, dGTP and dTTP) to synthesize DNA by extending the 3′-end of the primer. In each of four reactions, one per nucleotide, the corresponding dideoxynucleoside triphosphate (ddNTP) is also present. Incorporation of a ddNTP, rather than the corresponding dNTP, results in chain termination because the absence of a 3′-OH group prevents the formation of the next phosphodiester bond. So, for example in *Figure 5.3*, the A reaction contains the four dNTPs plus ddATP which acts as a terminator during DNA synthesis. As there is no 3′-OH group on the ddNTP it is not possible to form a phosphodiester bond so DNA synthesis of the growing DNA strand stops upon addition of ddATP. At each T position in the template there is a possibility that either dATP or ddATP will be added to the extending DNA chain. A small proportion of strands will terminate while the majority will continue being synthesized until the next T position where again a proportion will terminate by ddATP incorporation. Thus a series of fragments are generated that all start at the 5′-end of the primer and extend to one of the A positions in the growing chain, and thereby correspond to each T in the template strand. When these fragments are separated through a high-resolution denaturing polyacrylamide gel or capillary system they will migrate according to their length with the shortest fragments migrating fastest. This will create a ladder of fragments that represent the positions of each A in the synthesized DNA fragment. When the other reactions, C, G and T, are similarly performed using the same primer, template and the appropriate ddNTP, they also will produce a series of fragments terminating at the appropriate ddNTP. Comparing the migration rates of the fragments from the different reactions allows the sequence of the DNA to be read starting with the fastest migrating fragments that are closest to the primer.

In order to be able to read the reaction products they must be labeled in some way, usually by a radiolabel or a fluorescent label. Radiolabels are usually used for manual sequencing but the most common method for today involves the use of fluorescent dyes and automated detection systems. Two approaches are available; either primer-labeling or more commonly ddNTP labeling.

A variety of fluorescent dyes are available (e.g. JOE, ROX, FAM and TAMRA; Chapter 9) and can be linked to primers synthesized with a 5′-amino group, so that each fragment can be assigned to the corresponding nucleotide reaction by detection of a characteristic fluorescence wavelength. Primer labeling provides the highest quality and most uniform sequence data, however the dyes are commonly incorporated as ddNTPs (such as BigDye terminators™, Applied Biosystems). This allows any unlabeled primer to be used for sequencing, a particular advantage when using target sequence-specific primers rather than generic vector-specific primers. However, many universal, vector-specific fluorescently labeled primers are available commercially from several companies (Fluorescein labeled primers, Takara Mirus Bio; TAMRA labeled primers, USB Corp.).

Fluorescence detection systems include DNA sequencers based on slab gels (for example, ABI Prism™ 377 from PE Biosystems, ALF DNA Sequencer™ from Amersham/Pharmacia or the IR^2 from Li-Cor) or capillary electrophoresis systems (ABI Prism™ 3100 Genetic Analyzer or 3700 series from Applied Biosystems, Megabase from Molecular Dynamics or CEQ 2000 from Beckman-Coulter). These latter systems are based on DNA separation in thin-coated capillaries containing nonpolymerized gel matrices and laser detection systems. The introduction and removal of polymer from the capillaries, plus loading and running samples and fragment detection, are automated processes. Automated detection systems allow longer read lengths (800–1100 nts) than traditional radiolabeled approaches since the sample can be allowed to run for longer with real-time detection of fragments as they pass a laser and then continue to migrate into the lower buffer chamber (see below). In radiolabeled approaches the gel must be stopped and exposed to reveal the band pattern by autoradiography, thereby limiting the extent of sequence information (~300 nt) that can be detected.

Radioactive sequencing or the use of a single fluorophore requires the use of four separate sequencing reactions, one for each of the ddNTP terminators, and four lanes of a gel. In contrast, by using multiple dyes in dye primer reactions four separate sequencing reactions are required, but these can be mixed and loaded on a single gel lane or capillary. An advantage of multiple fluorescent dye terminators is that all four reactions can be performed in a single tube and loaded on a single gel lane or capillary. Only fragments that have incorporated a dideoxynucleotide will be dye-labeled and will be detected individually using a real-time laser gel scanner. This reduces the amount of work involved and avoids track-to-track variation during electrophoresis. Four laser systems allow up to 96 samples to be sequenced per slab gel. The larger capillary systems allow 96 samples to be sequenced every 1–2 h depending upon the amount of sequence data required per run.

Alternative fluorescence systems are available such as the IR^2 from Li-Cor where the four ddNTP-reactions are separated in adjacent lanes and detected by an infrared laser detection system. The output is an autoradiogram type image, but the fact that fragments are detected as they pass the laser system allows the gel to be run for longer. Together with its high sensitivity this allows detection of on average 1100 nucleotides from a single template. Since there are two fluorescent dyes that have nonoverlapping spectral features it is possible to mix two A reactions and separate them in one lane of the gel. Similarly two C, G and T reactions can be separated in the corresponding lanes. Simultaneous detection of the two fluorescent dyes allows up to 48 sequencing reactions to be separated on each gel.

Genome projects have significantly advanced the technologies associated with DNA sequencing including robotic automation of PCR set-ups, template purification, sequencing reactions and comb loading, all of which will serve to simplify sequencing of PCR products.

Primers for direct sequencing

The choice of primer for direct sequence analysis depends upon how much information is available before the PCR. One of the original PCR primers can be used, or ideally a nested primer that lies within the amplified fragment (Section 5.3). However, it is also possible to use generic sequencing primers such as M13 forward or reverse primers by including the appropriate sequences within the PCR primers when these were synthesized. This is particularly useful when a genomic PCR has been performed using degenerate primers and where there is no unique internal sequence information available for the amplified fragment. The inclusion of a generic primer site will ensure high-quality sequence information that would not be obtained by using the original degenerate mixture as sequencing primers. Examples of generic primer sites that can be added to PCR primers include:

- M13/pUC –47 sequencing primer (which also includes the –40 primer – shown in italics)
  5′-CGCCAGG*GTTTTCCCACTCACGAC*-3′;

- M13/pUC –20 sequencing primer
  5'-CGTTGTAAAACGACGGCCAGT-3';
- M13/pUC –48 reverse sequencing primer
  5'-AGCGGATAACAATTTCACACAGGA-3';]
- M13/pUC –24 reverse sequencing primer
  5'-AACAGCTATGACCATG-3';
- T7 promoter/universal primer
  5'-TAATACGACTCACTATAGGG-3';
- T3 promoter primer
  5'-ATTAACCCTCACTAAAGGGA-3';
- SP6 promoter primer (which includes SP6 universal primer –(italics)]
  5'-CATACG*ATTTAGGTGACACTATAG*-3'.

Isolating and purifying the PCR product

It is important to start the sequencing reaction with a pure template. A single PCR reaction should provide sufficient template for several DNA sequencing reactions. The methods used to isolate PCR fragments for direct sequencing (9) are identical to those used to isolate products before cloning and the range of options available is covered in more detail in Chapter 6. Usually an aliquot of the PCR is separated through an agarose gel. If there are several products, the band corresponding to the target fragment is excised and the DNA recovered by an appropriate method (Chapter 6). It may be necessary, depending upon yield of the product, to perform a preparative agarose gel with more of the original PCR product. An example of gel purification and cycle sequencing of the products is shown in *Figure 5.7*. Alternatively, if the analytical gel indicates a single product, it is possible to separate the DNA from other low-molecular-mass reaction components such as primers and nucleotides that may interfere with the sequencing, by using a simple spin column or other commercial PCR clean-up approach (Chapter 6). If the PCR product was biotinylated then a solid-phase system such as streptavidin-coated paramagnetic particles could be used (see below). The purified DNA is usually double-stranded unless it has been generated by asymmetric PCR (see below) and it is therefore necessary to denature the two strands and to prevent reannealing. This is achieved either by alkali denaturation followed by neutralization, or by heating to 100°C for 3–5 min and then snap-freezing in liquid nitrogen or dry-ice. Either a fluorescently labeled primer is annealed to the denatured template and standard dideoxy sequencing is performed, or an unlabeled primer is used with incorporation of fluorescent ddNTPs. A typical direct sequencing protocol is given in *Protocol 5.1*.

Generating single-stranded DNA templates

It is possible to generate single strand templates for DNA sequencing and other applications, for example as strand-specific probes. A simple approach is to use asymmetric PCR, in which one primer is added in vast excess (10–50-fold) over the other. During the first 20 or so cycles of PCR double-stranded product accumulates, but then with depletion of the low concentration primer the later cycles result in linear accumulation of one strand (*Figure 5.4*). This single-stranded product then provides a template for dideoxy sequencing (3,10). Although asymmetric PCR can be performed on any template source it is usually necessary to perform a titration to establish optimal primer:template ratios for each new template. A more efficient approach is to amplify a double-stranded PCR product and then to use an aliquot of this together with only one primer to generate a single-stranded product. It is important to remember that the primer for sequencing an asymmetric PCR product must be complementary to the asymmetrically amplified strand so you cannot use the same primer that was used for the asymmetric amplification.

Solid-phase sequencing

PCR products can be sequenced by capture of one strand onto a solid support, followed by alkaline denaturation and DNA sequencing (*Figure 5.5*). The most common method of immobilization is through the incorporation of biotin at the 5′-end of one primer. The resulting PCR product can then be isolated from solution by monodisperse paramagnetic beads coated with covalently bound streptavidin (11–13; *Figure 5.5*) or with a streptavidin affinity gel (14,15). Treatment with 0.1 M NaOH

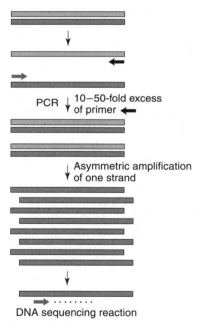

Figure 5.4

Asymmetric PCR for generating single-stranded template for sequencing. PCR is performed with an excess of one primer. When the low-concentration primer is exhausted the primer in excess continues to allow linear accumulation of one strand of DNA suitable for subsequent DNA sequencing.

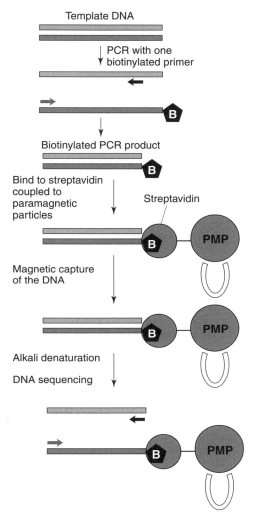

Figure 5.5

Solid-phase DNA sequencing. One of the PCR primers is biotinylated, allowing capture by streptavidin coupled to paramagnetic particles. This allows rapid purification of the DNA. Treatment with alkali leads to dissociation of the nonbiotinylated single strand. Both the strands can be used as templates for DNA sequencing reactions. The immobilized strand can be stored and reused for further reactions.

denatures the double-stranded captured product releasing the nonbiotinylated strand into solution and leaving the other strand attached to the matrix. The strand on the solid support can be resuspended in appropriate buffer and annealed with either a fluorescent labeled primer for dideoxy sequencing. Solid-phase sequencing allows both strands of the PCR product to be sequenced, either using alternate biotinylated primers, or by performing one dideoxy reaction in solid phase and the other in solution. The approach allows simple separation of the solid and solution phases for steps like buffer exchange and recovery of the template DNA. In addition the immobilized template can be reused for multiple sequencing reaction if necessary.

Cycle sequencing

Cycle sequencing involves incorporation of terminators during an asymmetric PCR. A thermal cycler (Chapter 3) is used to perform cycles of denaturation, annealing and extension on a template using a single primer resulting in linear amplification with simultaneous incorporation of dideoxynucleotide terminators. Although any form of template DNA can be used (double-stranded PCR products, plasmids, asymmetric PCR products), double-stranded PCR products or plasmids are most commonly used. For double-stranded molecules either strand can be sequenced depending on which primer is used. The cycle sequencing reaction products are analyzed using an appropriate gel or automated system with radioactive or fluorescent detection. A standard procedure is used for all templates in cycle sequencing and is superior to other PCR sequencing procedures in terms of length and accuracy of the sequence information obtained. Another benefit is that it can be used with relatively crude samples such as unpurified plasmid DNA from bacterial lysates because the linear amplification allows crude samples to be diluted to reduce the concentration of any inhibitory compounds that might normally poison the sequencing reaction. However, when dealing with difficult sequencing templates it is possible to add sequencing enhancers such as the SequenceRx Enhancer System from Invitrogen, containing seven co-solvents which are added individually to automated fluorescent DNA sequencing reactions to improve the length and quality of sequencing data.

Several thermostable DNA polymerases are available commercially in sequencing kits for cycle sequencing including AmpliTaq™ (Applied Biosystems), Vent$_R$®(exo⁻) (New England Biolabs) and Cyclist exo⁻ *Pfu* (Stratagene). Notably, AmpliTaq™ DNA polymerase FS (Fluorescent Sequencing) has become widely used for cycle sequencing, when modified to exhibit very little 5'→3' exonuclease activity and to incorporate ddNTPs much more efficiently than other enzymes. The ability to use lower concentrations of ddNTPs in both dye primer and dye terminator reactions significantly improves enzyme performance leading to higher sensitivity and therefore a requirement for reduced concentrations of template DNA and simpler dye removal before gel loading. Although most DNA sequencing companies offer services that include cycle sequencing a protocol for cycle sequencing using Applied Biosystems BigDye terminators is given in *Protocol 5.1*. The cycle parameters are significantly different to those used in standard PCR conditions in order to favor incorporation of ddNTPs. In particular the extension step involves an extended incubation at 60°C, and it is important to follow the recommended procedures provided by the manufacturers of the sequencing reagents to achieve reproducibly high quality results. An example of output from a cycle sequencing reaction on a PCR template is shown in *Figure 5.6*.

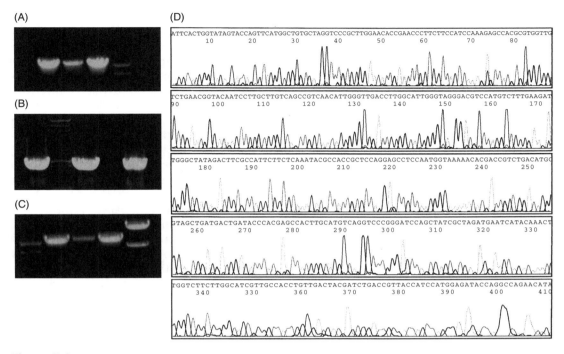

Figure 5.6

Direct sequencing of PCR products by cycle sequencing. The process was performed according to *Protocol 5.1*. (A)–(C), agarose gel analysis of: (A) an aliquot of the PCR products; (B) preparative gel after ethanol precipitation to concentrate the products (the DNA was recovered from gel slices using a gel purification kit); (C) aliquot of purified products to assess the amount of DNA to be used in each cycle sequencing reaction. (D) shows an example of DNA sequence from purified DNA with AmpliTaqFS and fluorescent-labeled ddNTPs with analysis on an ABI 373 DNA sequencer.

5.5 Direct labeling of PCR products and homogenous assays

Southern blot and dot blot hybridization procedures allow the simultaneous detection and verification of the identity of a PCR product. Once appropriate conditions for routine PCR product amplification are established it is often more appropriate to use more rapid and simpler approaches for detection of the amplified product. One approach is the direct labeling of the PCR product. It is possible to label the PCR products during the amplification reaction by the addition of labeled dNTPs or of radiolabeled, fluorescent or biotinylated primers that become incorporated into the PCR product. Various detection approaches can then be used, including gel electrophoresis followed by either X-ray film or phosphorimager exposure or by fluorescent detection on instruments such as the Applied Biosystems DNA sequencing instruments.

For some applications such as DNA footprinting and *in situ* PCR it is necessary to label the PCR product directly. Direct labeling of PCR products also increases the sensitivity of detection when low levels of amplification occur. It is simple and rapid compared with Southern blot and hybridization methods (Section 5.3), and a variety of labels can be incorporated. In direct labeling, one of the four nucleotides is either substituted for, or a reduced concentration of unlabeled nucleotide is supplemented with, labeled nucleotide. PCR is performed according to a standard protocol leading to direct incorporation of labeled nucleotides into the final PCR product. Radiolabelled nucleotides are frequently used when analyzing low levels of amplification with detection by autoradiography or phosphorimager analysis of a gel or by scintillation counting of a trichloroacetic acid (TCA)-precipitated aliquot. Care must be taken to avoid excessive exposure to the radioactive source and the thermal cycler should always be shielded from the user. Nonradioactive labels have become increasingly popular as alternatives to radiolabels although in some cases they are not so sensitive. These include hapten-conjugated nucleotides such as biotin-dUTP and digoxigenin (DIG)-dUTP and fluorescent nucleotides such as fluorescein-dUTP. An alternative to direct incorporation into the PCR product is through the use of labeled primers. Although these can be significantly more expensive than nonlabeled primers they allow single additions of a label per product molecule allowing more accurate quantitation and are very useful for high-throughput procedures. Most methods for real-time detection of PCR utilize labeled PCR primers as described in Chapter 9. A useful discussion of the relative sensitivity of radioactive versus nonradioactive labels can be found in Mundy *et al.* (16).

Scintillation proximity assay

The scintillation proximity assay (SPA) (17–19) is a methodology available from Amersham Biosciences. SPA beads contain a fluor and are coated with appropriate acceptor molecules that bind radiolabeled ligands in solution. The ligand is isotopically labeled with a low-energy radiation emitter such as tritium (^{3}H) whose energy is dissipated in the aqueous medium. However, upon binding of the ligand to the SPA bead the label will be sufficiently close to activate the fluor and generate light that can be detected by a scintillation counter. In practice the SPA beads are coated with streptavidin that binds biotin. If one of the PCR primers is biotinylated then during amplification in the presence of [^{3}H] dTTP the products will become both radiolabeled and biotinylated. An aliquot can then be added to SPA beads and the amount of product quantitated by the scintillation signal resulting from streptavidin binding of PCR product to the SPA bead. The process can be made selective for a particular PCR product by using nonbiotinylated PCR primers during the synthesis of labeled product. The products are then heat denatured and mixed with a specific 5′-biotinylated oligonucleotide probe that will hybridize to the target PCR product allowing detection of this product but not other products that may also be present in the reaction (*Figure 5.7*). Unbound labeled ligands in solution are too distant from the SPA beads to facilitate energy transfer to the fluor and therefore the signal-to-noise ratio is low and there is a linear response over two orders of magnitude.

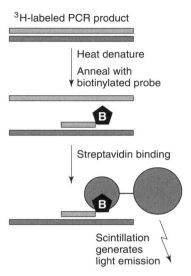

Figure 5.7

Scintillation proximity assay. A biotinylated oligonucleotide probe anneals to the target PCR product, allowing its capture by streptavidin immobilized on a fluor-containing bead. The proximity of the ³H-labelled PCR product to the fluor leads to scintillation that can be measured.

Immunological detection of PCR products

Enzyme-linked immunosorbent assay (ELISA) methods can be used to detect PCR products. These approaches offer high sensitivity allowing detection of products at levels of around 10 pg. One approach (Hybrid Capture® 2 Technology, Digene Diagnostics Inc.) involved the use of a monoclonal antibody that detects RNA:DNA hybrids (20,21). PCR amplification uses one biotinylated and one nonbiotinylated primer, the product is heat denatured and hybridized in solution to a complementary unlabeled RNA probe. The resulting DNA:RNA hybrids are isolated from solution by biotin binding to streptavidin-coated wells in a microtiter plate. The monoclonal antibody conjugated to alkaline phosphatase is added to bind to the trapped DNA:RNA hybrids and detected by adding a colorimetric substrate. A similar DNA Enzyme Immunoassay System (Sorin Biomedica now DiaSorin), is based on a biotinylated probe immobilized in the wells of a streptavidin-coated microtiter plate. Denatured PCR products are allowed to hybridize to the probe, creating double-stranded DNA segments that can be detected using an anti-double-stranded DNA antibody with a coupled enzyme reaction using a chromogenic substrate. This approach has been used to detect various genetic lesions, virus and parasite sequences.

5.6 *In vitro* expression of PCR products

As described in Chapter 3, in addition to including promoter sequences as 5′-additions to PCR primers, other gene expression signals can be added.

For example, inclusion of a translation initiation region downstream (3′) to a phage promoter will yield a transcript that can subsequently be translated *in vitro*. An *E. coli* S30 extract has been used successfully when prokaryotic translation signals are included. For a eukaryotic system such as a wheat germ or rabbit reticulocyte system, eukaryotic translational signals are necessary and the transcript should be synthesized in the presence of 7mGpppG, so that a 5′-capped RNA is produced. This approach (*Figure 5.7*) is known as RNA amplification with *in vitro* translation (RAWIT; 22).

It is now a relatively easy process to amplify a coding region of a gene or cDNA and to use the PCR product as a template for direct expression of the encoded protein. Generally the amounts of protein expressed are relatively small so it is common to include an appropriate detection system such as a radiolabeled amino acid during protein synthesis coupled with autoradiography of a polyacrylamide gel or by adding a tag such as poly-histidine (His$_6$-tag) during the PCR for subsequent Western blot detection using an anti-His tag antibody. As well as complete coding regions, PCR variants such as truncated fragments can readily be analyzed. For example, the expression of a series of PCR fragments derived from a gene, with subsequent Western blot detection using a monoclonal antibody, could be used for epitope mapping of an antigenic determinant in the protein. Similarly, the addition of a His$_6$-tag during PCR amplification to mutated versions of a protein (removal of silent codons) followed by antibody detection has been used for high-throughput screening strategies to identify protein versions that express to high levels (23). By using such an approach several mutants can be analyzed for expression levels within a day followed by selection for future heterologous expression strategies.

The *in vitro* expression of proteins can also be used for functional studies of proteins. For example, enzymes synthesized *in vitro* are likely to be correctly folded and to retain enzymatic activity which can subsequently be measured in an appropriate assay system. The approach could also be used for rapid functional domain mapping of proteins where fragments that represent protein domains are expressed and functionally analyzed using *in vitro* assays. Furthermore, rapid *in vitro* protein–protein interaction studies could be performed using this technique where not only full-length version of proteins could be expressed and analyzed but also protein fragments for mapping of interaction domains.

The use of *in vitro* expression approaches to analyze protein products is likely to gain momentum for the analysis of known proteins as well as for studies on unknown proteins encoded by ORFs (open reading frames) generated by genome-sequencing projects. In this era of genome sequencing there is an increasing demand to assign functions to ORFs and the *in vitro* expression approach will certainly be of benefit.

Efficient expression obviously relies upon the presence of appropriate signals for both transcription and translation and these vary depending upon whether a prokaryotic or eukaryotic coupled expression system is to be used. A number of commercially available transcription and translation systems are available for *in vitro* expression studies of PCR products. In general the promoter used is a phage promoter such as T7, T3 or SP6 that is recognized by the corresponding phage RNA polymerase (*Figure 5.8*).

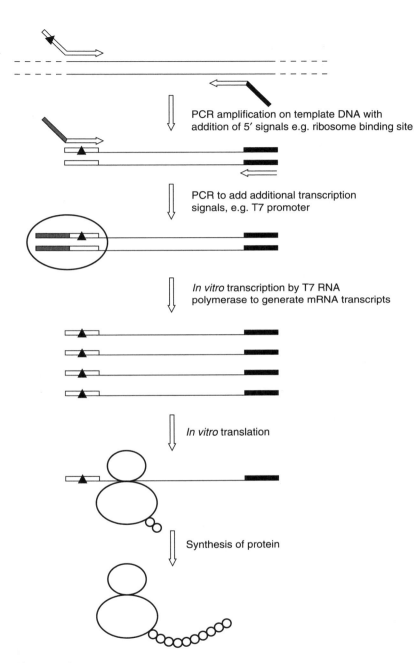

Figure 5.8

RNA amplification with *in vitro* translation. Two rounds of PCR are used to add regulatory signals to the cDNA template. ▲, ribosome binding site; ▬ T7 transcription terminator; ▬ T7 promoter. The PCR product is then used in a transcription and translation reaction to produce the protein product.

It is possible to design PCR primers that have 5'-tails corresponding to a promoter and translation initiation sequence. For example the sequences:

- GAATTCTAATACGACTCACTATA<u>GGGGTTAACTTTAAGAAGGAGATATA CATATG</u> [1]
- CCAAGCTTCTAATACGACTCACTATA<u>GGGTTTTTATTTTTAATTTTCTTT CAAATACTTCCACC**ATG**</u> [2]

could be used as 5'-extensions to primers to add a T7 RNA polymerase sequence and untranslated leader to PCR products for expression in either [1] an *E. coli* S30 or [2] a wheat germ system, respectively (24). The transcribed region is underlined and the initiating ATG start codon is in bold. An alternative promoter, such as SP6, can also be used in place of T7. This primer-tailing approach requires the synthesis of long primers 70–80 nucleotides in length for each PCR product to be analyzed. An alternative approach is to incorporate these sequences in a second PCR following ligation to a vector. Various manufacturers provide suitable systems, for example Roche or Invitrogen.

References

1. Southern EM (1975) Detection of specific sequences among DNA fragments separated by gel electrophoresis. *J Mol Biol* **98**: 503–517.
2. Wong C, Dowling CE, Saiki RK, Higuchi RG, Erlich HA, Kazazian HH Jr (1987) Characterization of β-thalassaemia mutations using direct genomic sequencing of amplified single copy DNA. *Nature* **330**: 384–386.
3. Wrischnik LA, Higuchi RG, Stoneking M, Erlich HA, Arnheim N, Wilson AC (1987) Length mutations in human mitochondrial DNA: direct sequencing of enzymatically amplified DNA. *Nucleic Acids Res* **15**: 529–542.
4. Fattorini P, Ciofuli R, Cossutta F, Giulianini P, Edomi P, Furlanut M, Previdere C (1999) Fidelity of polymerase chain reaction-direct sequencing analysis of damaged forensic samples. *Electrophoresis* **20**: 3349–3357.
5. Sanger F, Nicklen S, Coulson AR (1977) DNA sequencing with chain-terminating inhibitors. *Proc Natl Acad Sci USA* **74**: 5463–5467.
6. Tabor S, Richardson CC (1987) DNA sequence analysis with a modified bacteriophage T7 DNA polymerase. *Proc Natl Acad Sci USA* **84**: 4767–4771.
7. Innis MA, Myambo KB, Gelfand DH, Brow MA (1988) DNA sequencing with *Thermus aquaticus* DNA polymerase and direct sequencing of polymerase chain reaction-amplified DNA. *Proc Natl Acad Sci USA* **85**: 9436–9440.
8. Mead DA, McClary JA, Luckey JA, Kostichka AJ, Witney FR, Smith LM (1991) *Bst* DNA polymerase permits rapid sequence analysis from nanogram amounts of template. *BioTechniques* **11**: 76–78.
9. Vogelstein B, Gillespie D (1979) Preparative and analytical purification of DNA from agarose. *Proc Natl Acad Sci USA* **76**: 615–619.
10. Gyllensten UB, Erlich HA (1988) Generation of single-stranded DNA by the polymerase chain reaction and its application to direct sequencing of the HLA-DQA locus. *Proc Natl Acad Sci USA* **85**: 7652–7656.
11. Hultman T, Bergh S, Moks T, Uhlen M (1991) Bidirectional solid-phase sequencing of in vitro-amplified plasmid DNA. *BioTechniques* **10**: 84–93.
12. Hultman T, Stahl S, Hornes E, Uhlen M (1989) Direct solid phase sequencing of genomic and plasmid DNA using magnetic beads as solid support. *Nucleic Acids Res* **17**: 4937–4946.
13. Kaneoka H, Lee DR, Hsu KC, Sharp GC, Hoffman RW (1991) Solid-phase direct

DNA sequencing of allele-specific polymerase chain reaction-amplified HLA-DR genes. *BioTechniques* **10**: 30.

14. Stahl S, Hultman T, Olsson A, Moks T, Uhlen M (1988) Solid phase DNA sequencing using the biotin-avidin system. *Nucleic Acids Res* **16**: 3025–3038.

15. Mitchell LG, Merril CR (1989) Affinity generation of single-stranded DNA for dideoxy sequencing following the polymerase chain reaction. *Anal Biochem* **178**: 239–242.

16. Mundy CR, Cunningham MW, Read CA (1991) Nucleic acid labelling and detection. In Brown TA (ed) *Essential Molecular Biology II: A Practical Approach*, pp. 57–109. Oxford University Press, Oxford, UK.

17. Hart HE, Greenwald EB (1979) Scintillation proximity assay (SPA) – a new method of immunoassay. Direct and inhibition mode detection with human albumin and rabbit antihuman albumin. *Mol Immunol* **16**: 265–267.

18. Udenfriend S, Gerber L, Nelson N (1987) Scintillation proximity assay: a sensitive and continuous isotopic method for monitoring ligand/receptor and antigen/antibody interactions. *Anal Biochem* **161**: 494–500.

19. Gyllensten UB (1989) Direct sequencing of *in vitro* amplified products. In Erlich HA (ed) *PCR Technology*, pp. 45–60. Stockton Press, New York.

20. Bobo L, Coutlee F, Yolken RH, Quinn T, Viscidi RP (1991) Diagnosis of *Chlamydia trachomatis* cervical infection by detection of amplified DNA with an enzyme immunoassay. *J Clin Microbiol* **29**: 2912.

21. Lazar JG (1993) A rapid and specific method for the detection of PCR products. *Am Biotechnol Lab* **11**: 14.

22. Sarkar G, Sommer SS (1989) Access to a messenger RNA sequence or its protein product is not limited by tissue or species specificity. *Science* **244**: 331–334.

23. Betton JM (2004) High throughput cloning and expression strategies for protein production. *Biochimie* **86**: 601–605.

24. Lesley SA (1995) In McPherson MJ, Hames BD, Taylor GR (eds) *PCR 2: A Practical Approach*. IRL Press at Oxford University Press, Oxford, UK.

Protocol 5.1 Cycle sequencing – Applied Biosystems Big Dye terminators

EQUIPMENT

Agarose gel apparatus

Microcentrifuge

Thermal cycler

0.5 ml microcentrifuge tubes

or 96 well plate

MATERIALS

Agarose gel

Gel loading buffer

PCR purification or gel purification kit

Molecular size markers

Sequencing premix (containing buffer, dNTPs, dye terminators, AmpliTaqFS)

Sequencing primer (2 pmol μl^{-1})

Mineral oil (if no heated lid)

Phenol/chloroform from PE Biosystems at room temperature (it is critical that this is of the highest quality to avoid fluorescent contaminants that otherwise affect the sequencing gel)

3 M sodium acetate, pH 5.2

125 mM EDTA

Ethanol, ice-cold 95% and 70%

1. Purify the PCR product either by use of a PCR clean-up kit or by isolation from an agarose gel (Chapter 6). Resuspend the DNA in a small volume, <30μl, of elution buffer.

2. Analyze 0.1 vol of the DNA on an agarose gel containing 1 μg ml^{-1} ethidium bromide, alongside a known quantity of molecular size markers. Estimate the quantity of sample DNA by comparing the relative intensity of the ethidium bromide-stained band with the marker bands. A total of 30–90 ng of DNA is recommended for direct sequencing of PCR products.

3. Combine in a 0.5 ml microcentrifuge tube:
   - 30–90 ng template DNA;
   - 1.6 µl (3.2 pmol) primer;
   - 1 µl sequencing premix;
   - × µl distilled water to give a total volume of 20 µl.

   If necessary overlay the tubes with a drop of mineral oil.

4. Place in a thermal cycler incubate for 1 min at 96°C then subject to 25 cycles of the following:
   - 96°C, 10 s;
   - 50°C, 5 s;
   - 60°C, 4 min.

   This takes approximately 2 h to complete.

5. Add 2 µl sterile 3 M sodium acetate, 2 µl 125 mM EDTA vortex for 1 s, then add 50 µl ice-cold 95% ethanol, vortex for 1 s and leave on ice for up to 15 min.

6. Pellet the DNA in a microcentrifuge (13 000 *g*), preferably at 4°C for 15 min. As the pellet may not be visible (unless you use a pellet paint, Chapter 3) orient the tubes conveniently with the hinges to the outside to indicate the side of the tube on which the DNA will be pelleted.

7. Remove the ethanol by aspirating off with a pipettor, taking care not the touch the side of the tube with the DNA.

8. Wash the pellet by adding 100 µl ice-cold 70% ethanol, vortex briefly and centrifuge as before with the tube in the same orientation.

9. Aspirate off the ethanol, gently tap the mouth of the tube onto some tissue to remove most of the residual ethanol and dry briefly (up to 5 min) in a centrifugal evaporator to remove traces of ethanol. Do not overdry the DNA as this makes redissolving difficult.

10. The samples are now ready to be resuspended in dye-formamide and fractionated on an ABI sequencing instrument.

Purification and cloning of PCR products

6

6.1 Introduction

Once you have generated a PCR product it must often be cloned to provide a permanent source of the amplified DNA fragment(s) for future use. This Chapter outlines methods for purifying PCR products prior to cloning or direct sequence analysis (Chapter 5), or for use as hybridization probes, and then describes strategies for cloning PCR products into appropriate vectors.

PCR is a superb technique for the isolation of a target DNA sequence from either genomic DNA or cDNA in a relatively short time, avoiding many of the time-consuming aspects of 'traditional' gene cloning procedures. However, once you have your product you will often clone it into a suitable vector to provide a ready supply of the DNA without the need to repeatedly amplify the product from its original source. This will allow you to use the product for a variety of purposes, either as control DNA in subsequent experiments or for further detailed investigation.

A critical step in planning a PCR experiment is to consider the vector and cloning strategy that you will adopt *before* you design and order the primers to perform the PCRs. Although, as we will discuss, it is possible to clone any PCR product, it is most efficient to design the experiment first in order to optimize primer design and build appropriate features into the primers before PCR. For example, these may include suitable restriction sites, regulatory elements such as promoters, or additional nucleotides to encode a peptide linker or to ensure the reading frame of a coding region is maintained. It is this ability to tailor make primers with the most appropriate features and thus to modify the resulting PCR product that makes PCR such a powerful method compared with traditional 'cut-and-paste' experiments based on naturally occurring restriction enzyme sites. Once the primers have been designed and the PCR product has been generated you will invest significant time and effort in cloning and characterizing your amplified DNA. It can be very helpful to perform your experiments *in silico* first to ensure correct design features are considered such as maintenance of open reading frames. An appropriate software system can be used, such as Vector NTI (Invitrogen; see http://www.invitrogen.com/, where a limited number of tools are available freely online). Take note of the methods described in Chapter 5 that deal with verification of the PCR product and make sure that it is the correct one either before you clone it, or as the first analysis of resulting clones.

6.2 Purification of PCR products

Advantages of purifying PCR products for sequencing or cloning include removal of:

- primers, nucleotides and buffer components;
- nontarget amplification products;
- compounds that may inhibit the ligation reaction.

In addition the concentration of product can be increased.

The major disadvantage is loss of product as no purification procedure has a 100% recovery rate. Clearly the advantages outweigh the disadvantages and so it is strongly recommended that PCR products are purified prior to the ligation reaction. There are a range of alternative protocols for product purification depending upon the efficiency and specificity of the PCR and the subsequent use for the purified DNA. The following sections do not attempt to provide a comprehensive list of available methods and commercial kits but do describe the main approaches and their principles.

Commercial DNA purification kits

The simplest, most convenient and most reproducible approach is to use a commercial PCR purification kit. Such kits for purification of PCR-generated DNA fragments, either from solution or from a gel slice, are available from most large molecular biology reagent supply companies, and in general they perform equally well. Most kits are based on the retention of DNA fragments of greater than around 100 bp on some form of solid support, such as a silica membrane. Following washing steps to remove dNTPs, buffer and unincorporated primers, a final elution step allows recovery of the bound DNA in a reasonably small volume. The benefit of such systems is that they remove the need for steps such as phenol extractions and ethanol precipitations and they are relatively quick and easy. For example, the QIAGEN QIAquick Gel extraction kit, like many commercial kits, is based on spin-column technology together with absorption of DNA to a membrane. It can be used for either gel extraction or direct purification of PCR products. For gel extractions, the PCR products should be size-fractionated through an agarose gel and the DNA band of interest cut from the gel in the smallest possible volume of agarose, under UV illumination, using a fresh razor blade. Remember to take precautions such as wearing gloves and a face shield to prevent UV irradiation damage. Carefully trim away as much excess agarose as possible. The gel slice is melted in the presence of a chaotropic salt such as sodium iodide followed by absorption to a membrane in a spin column. The bound DNA is then washed, removing contaminants, followed by elution into a Tris-based buffer. The same procedure is used for post-PCR clean-up without gel separation. In this case the contents of the PCR tube are mixed with a high salt solution, loaded onto the spin column, which is washed to remove dNTPs and primers, and the PCR products are eluted. The procedure is rapid (~15 min) and results in highly purified DNA for use in ligation reactions. Other similar kits are available from a range of molecular biology suppliers.

A benefit of commercial kits is that they generally avoid ethanol precipitation steps, although for small amounts of product elution in the recommended volume of around 30–50 μl of buffer may lead to samples being too dilute. However, as they are in water or a low-salt buffer the sample can usually be concentrated by evaporation for a few minutes in a spin-vac to achieve the desired concentration.

Ethanol precipitation

Ethanol precipitation can be used as a fairly crude purification tool for removal of nucleotides and salts, with the added benefit that it also concentrates DNA samples. It can remove short oligonucleotides (<15 nucleotides) but can lead to coprecipitation of the longer oligonucleotides used as primers in many PCR applications. Several new approaches for recovery of DNA do not require concentration of DNA by precipitation from ethanol. The main justification for including some discussion of the method here is that ethanol precipitation is a simple, cheap and well-tested tool. The DNA solution is increased in salt concentration and precipitated by addition of ethanol. Traditionally ethanol precipitation was performed at low temperature, usually by incubating at –20°C or –70°C, however, it is now recognized that this results in increased precipitation of salt and so incubation at room temperature or in an ice bucket is now more common. After collecting the DNA by centrifugation, usually at 13 000 g in a microcentrifuge, the pellet is washed in 70% ethanol to remove excess salt before briefly drying and redissolving in an appropriate buffer such as 10 mM Tris-HCl (pH 8.0), 1 mM EDTA. Particularly where small quantities of DNA are being precipitated it can be difficult to see a pellet. In such cases the microcentrifuge tubes should be placed in the microcentrifuge in a defined orientation, for example with the hinge upwards, so the position of the pellet can be identified even if it is not visible. An inert carrier compound such as glycogen used at 50–150 µg ml^{-1} or linear acrylamide used at 10–20 µg ml^{-1} such as those from Ambion can be added to increase the size of the pellet. There are also now colored carriers available. Examples include Glycoblue, a derivatized glycogen (Ambion) and Pellet Paint™ coprecipitant (Novagen) that is available either in a fluorescent or nonfluorescent (NF) format. These reagents provide a visual indicator of the presence and position of an ethanol pellet. The Pellet Paint™ NF reagent is compatible with preparation of samples for fluorescent sequencing applications involving dyes such as the BigDye™ terminators where the fluorescent coprecipitant would interfere with sequence detection.

In-gel ligation

High-purity, low-melting-temperature agarose does not inhibit DNA ligase activity. Thus, PCR products can be cloned directly after agarose gel electrophoresis and without recovery from the agarose. The PCR products should be size fractionated through a low-melting-temperature agarose gel and the DNA band of interest cut from the gel as described above. Ideally the final concentration of agarose should be 0.4% or lower so if, for example, you used a 4% gel then the agarose can only comprise 0.1 vol. of the ligation reaction volume. The gel slice should be equilibrated with water in a microcentrifuge tube for about 30 min to remove the electrophoresis buffer. The agarose slice is melted by heating to 50°C. The ligation reaction components are set up as for a standard reaction, with the exception of the PCR product. The appropriate volume of gel, held at 50°C is then pipetted into the ligation mix held at 37°C, and mixed immediately. This ensures the agarose does not set on contact with the other reaction components. The ligation reaction can then be incubated at the desired temperature (15–37°C). Depending on the temperature at which the ligation reaction is performed the agarose may partly solidify, but this is not a problem. Advantages of in-gel ligation are that it is rapid and relatively cheap (although high-quality agarose is expensive) and DNA loss is avoided. As a comment on any gel purification procedure, even when well-separated bands are purified from a gel, there can be some cross-contamination from other DNA molecules. This makes it important to confirm the identity of an insert in clones derived from the isolated DNA. If the PCR product is to be used for direct analysis such as DNA sequencing without cloning, then any such minor cross-contamination will not be an issue.

Spin columns

An early method of DNA purification used a siliconized glass wool plug and standard microcentrifuge tubes. The principle of the technique is that the glass wool physically retains an agarose slice while under centrifugal force the buffer and DNA are forced out of the gel and can pass through the glass wool plug. A small plug of siliconized glass wool is placed in the bottom of a 0.5 ml microcentrifuge tube containing a pin-sized hole in the bottom to allow liquid to pass through during centrifugation. The gel slice containing DNA is placed on top of the glass wool cushion and the tube placed inside a 1.5 ml microcentrifuge tube before centrifugation at 13 000 g for 1–2 min. The liquid in the larger tube should contain DNA from the gel and can be concentrated by ethanol precipitation if necessary. Once it is placed in the tip of filter unit the gel slice can be subjected to a freeze–thaw cycle by placing the unit at –20°C until the gel is frozen and then allowing it to thaw at room temperature. Such a treatment can increase the recovery of product. This method is cheap and generally reliable although occasionally agarose components can pass through the glass wool plug. It is better to use a standard agarose rather than a low-melting-temperature agarose as the former will be less likely to disintegrate during centrifugation. Generally this approach has been superseded by commercially available spin filters such as 0.22 μm Costar® Spin-X® centrifuge tube filters (Corning Life Sciences). The agarose slice is simply placed in the filter and centrifuged in a microcentrifuge for 1–2 min. Quantitative recovery of product depends on the size and amount of DNA being purified, and is generally more efficient for shorter fragments. In general recovery is usually less than 50%.

Electroelution

After agarose gel electrophoresis the DNA band is cut from the gel and the DNA is eluted from the gel slice by means of an electrical current. There are many approaches to electroelution and we mention only two here. The first approach does not require special apparatus. The gel slice is placed at one side of a piece of preboiled (1) dialysis tubing that also contains a small volume (100–500 μl depending on the gel slice size) of the agarose gel running buffer. After sealing, the dialysis tube is placed in the gel electrophoresis tank containing the same buffer as in the tubing and gel slice. The gel slice is closest to the anode (–) and an electrical current is applied to electrophoretically elute the DNA from the gel slice so it becomes trapped on the surface of the dialysis tubing. It is recommended that the elution be allowed to proceed for 30 min at 50–75 V when using a mini-gel apparatus. DNA elution from gel slices can be monitored by use of a hand-held UV lamp (365 nm) to visualize ethidium bromide fluorescence. Often the DNA will accumulate on the cathode-facing inner surface of the dialysis membrane. It can be released by reversing the current in the electrophoresis tank for 30–60 s. Alternatively, following removal of the gel slice, it can be released back into solution by gentle agitation or pipetting of the solution against the membrane. It can sometimes be convenient to remove some of the buffer from the dialysis bag before dislodging the DNA to allow a more concentrated solution to be recovered. If necessary the DNA can be concentrated by ethanol precipitation (see above). There are also various commercial apparatus for electroelution, for example from Stratagene and Millipore.

Silica matrix or Geneclean purification

This approach is the basis for most commercial purification kits. It is based on the observation that DNA could be released from agarose gel slices and bound to silica particles in the presence of a chaotropic salt (2,3). A gel slice containing DNA is excised from an agarose gel and allowed to dissolve in 1 ml of 6 M sodium iodide at 55°C. Once dissolved around 10 μl of a silica fine-particle suspension is added, mixed and incubated with constant but gentle shaking for 10 min. The silica fines bind the DNA that can be collected by microcentrifugation followed by washing three times with 70% ethanol. The pellet is air-dried briefly (about 5 min) and the DNA eluted in 20 μl of water by incubation at 45°C for 1 min. Although this approach results in good recovery (up to 80%) of DNA from agarose gels it is not recommended for large DNA fragments (10–15 kbp) as shearing is often observed. For example, Geneclean can be obtained from Q-biogene.

6.3 Introduction to cloning of PCR products

The success of cloning PCR-generated fragments depends on several factors, including PCR product purity (Section 6.2), the choice of restriction enzyme(s), primer design and the plasmid you choose to use as the recipient vector. Although the cloning of PCR-amplified products can sometimes prove difficult, new and improved vectors and procedures have been developed to increase cloning efficiency. The following sections describe factors that should be considered in order to successfully clone your PCR product and will outline several ways to increase your PCR cloning efficiency.

PCR re-amplification

Occasionally you will have a low yield of PCR product. To increase the yield it is possible to re-amplify using PCR. Essentially a small aliquot of the products of the first PCR is used as template in a second round of PCR using the same primers and reaction conditions. In this case there is no need to purify the products of the first PCR before performing the second PCR, simply use a 1–5 µl aliquot of the first PCR reaction mix as the template for the second PCR. Of course one potential disadvantage of the increased number of PCR cycles is the increased possibility of accumulating PCR-mediated mutations in the final PCR products. Use of 'proofreading' DNA polymerases (Chapter 3) reduces, but does not eliminate this possibility. Generally therefore, PCR re-amplification should be avoided as a routine procedure to increase product yield. Rather, it is more appropriate either to increase the amount of template used, or to perform several identical PCR amplifications using a standard number of cycles (25–30 cycles) and to pool the products. Nonetheless, it may be appropriate to use a re-amplification step if there is negligible product visible and you suspect, for example, that either the amount of starting template was very low, or the reaction has not worked efficiently, perhaps due to some contaminant. The effect of performing a further PCR would be to use the enriched template preparation to amplify the product sufficiently to visualize it, or to dilute out contaminants interfering with the reaction. If products from re-amplifications or nested PCRs are to be cloned it is important to ensure that several independent clones are sequenced to identify those containing the correct sequence and to discard any that may contain a mutation. This is obviously more difficult for clones whose sequence is not already known and in such cases may require the sequencing of 10–12 clones to identify the consensus.

Why can PCR cloning be a problem?

You may have heard that the efficiency of PCR cloning can be low, but careful experimental design can reduce such difficulties. One important source of difficulty is the terminal transferase activity of *Taq* DNA polymerase that leads to the addition of an additional nucleotide, usually an A, at the 3'-end of the newly synthesized DNA strand. This non-template-directed addition leads to PCR products that do not have blunt ends as expected, but rather have single nucleotide extensions. This phenomenon explains the inefficiency of blunt-end ligations involving PCR products. In order to generate blunt-end PCR fragments it is necessary to treat the DNA with a proofreading enzyme such as the Klenow fragment of DNA polymerase I, or T4 or T7 DNA polymerase, or a proofreading thermostable DNA polymerase, in the presence of the four dNTPs (see *Protocol 6.1*). This procedure results in the enzyme removing the unpaired terminal nucleotide, but the presence of the dNTPs means that if the enzyme removes the next nucleotide this is immediately replaced by its 5'→3' DNA synthesis activity, leaving a blunt or 'polished' end. The 'terminal A' issue does not generally occur when a thermostable proof-reading DNA polymerase is used as these enzymes would remove any unpaired nucleotide they erroneously added. Several commercial systems are available for cloning PCR products by exploiting the additional A added by *Taq* DNA polymerase.

6.4   Approaches to cloning PCR products

Essentially any cloning vector can be used for cloning a PCR product, although as with any cloning experiment success is often better with relatively small vectors (2.5–5 kbp). An increasing range of vectors are available from molecular biology reagent suppliers that:

- allow cloning of restriction digested PCR products;
- allow efficient blunt-end cloning of proofreading enzyme products;
- exploit the additional A on *Taq* PCR products (4); or
- utilize topoisomerase-mediated (TOPO) ligation (5) for very rapid (5 min) cloning reactions;
- exploit the addition of 5′-sequences on primers to allow ligation-independent cloning or recombinational cloning.

Various approaches for cloning PCR products are outlined below, and the features of the PCR product and vector are summarized in *Table 6.1*.

Restriction enzyme cloning

It is common to incorporate restriction enzyme sites into the primers used to generate the PCR products (6). When the PCR product is digested with these restriction enzymes the resulting fragment can be ligated with a suitably restricted vector molecule. Often it is convenient to introduce different restriction enzyme sites at the two ends of the PCR product to allow directional cloning into the doubly digested vector, with at least one of the enzymes generating a cohesive or 'sticky' end (*Table 6.2*). This double-digest strategy can also avoid the need to use alkaline phosphatase to dephosphorylate the vector, a step that is necessary to prevent religation of the vector alone if it is restricted with a single enzyme. The introduction of restriction sites into the primer is straightforward (Chapter 3) and there are two approaches. Most commonly the site is added as a 5′-extension to the PCR primer (*Table 6.2*), or if there is a sequence within the PCR primer that differs by only one or two nucleotides from a restriction enzyme site, these nucleotides can be changed or mutated to generate the new restriction site within the original sequence. There are some issues that must be considered when designing such primers. The positioning of the restriction site in relation to the 5′-end of the primer and the enzyme you choose dictate the efficiency of digestion and the overall success of your cloning experiment (7). A useful source of information about how many nucleotides to add to the 5′-end of primers for digestion by different enzymes is given in an Appendix to the New England Biolabs molecular biology products catalogue. It is recommended that between 3 and 10 nucleotides should precede the restriction enzyme site in order to ensure efficient cleavage of the site within the terminus of a PCR product (*Table 6.2*). It is best to err on the side of caution and add sufficient overhang nucleotides since the cost of additional nucleotides added to a primer sequence is more effective than having to adopt some alternative strategy to ensure efficient restriction enzyme cleavage.

   If this issue does prove problematic, one approach that has been reported to overcome some difficulties with restriction enzyme digestion is to blunt-

Table 6.1 Approaches to cloning PCR products and the vector features necessary for different PCR cloning strategies

Cloning strategy	Type of end	Sequence of end	Vector properties
Taq DNA polymerase	3'-dA overhang	5' NNNNNN......3' 3' ANNNNNN......5'	TA vector + ligase TOPO TA vector
Proofreading DNA polymerase	Blunt end	5' NNNNNN......3' 3' NNNNNN......5'	Blunt ended vector, e.g. SmaI digested (CCC/GGG) then alkaline phosphatase then ligase
Added restriction site (blunt end)			Zero Blunt® Zero Blunt® TOPO
Added restriction site (cohesive end)	5' or 3' overhang e.g. *EcoRI* *BamHI* *PstI*	5'AATTCNNNN......3' 3'GNNNN......5' 5'GATCCNNNN......3' 3'GNNNN......5' 5'GNNNN......3' 3'ACGTCNNNN......5'	Similarly digested vector If single digest, then treat vector with alkaline phosphatase then ligase If double digest, add ligase
Directional TOPO cloning	Add the appropriate sequence to the 5' end of the sense primer	5'CACC-target sequence......3' 3'GTGG-target sequence......5'	TOPO activated vector 5'......vector3' 3'......vectorGTGG5' Strand invasion of the added sequence by the complementary vector tail and TOPO ligation. The other end of the product is joined to the vector by a blunt end TOPO reaction

Table 6.1 *continued*

Cloning strategy	Type of end	Sequence of end	Vector properties
Ligation independent cloning	Polynucleotide tail added by Terminal deoxynucleotidyl transferase (TdT)		Restricted vector treated with (TdT) and complementary nucleotide
		`5'GGGGGGGGGGGG...3'` `3'...5'`	`5'...3'` `3'...CCCCCCCCCCCCC...5'`
Ligation independent cloning	Specific sequence added to 5' end of upstream primer	`5'GAC GAC GAC AAG ATX-targetsequence 3'` `3' AX-targetsequence 5'`	`5'vector......3'` `3'vector......CTGCTGCTGTTCT5'`

	Specific sequence added to 5' end of downstream primer	`5'target sequence-A`
		`3'target sequence-TGG CCC GAA GAG GAG5'`
			`5'CC GGG CTT CTC CTC......vector3'`
			`3'......vector5'`
			Anneal vector and insert, 22°C, 5 min

	Then treat PCR product with T4 DNA polymerase + dATP		
Gateway cloning	Add attB sites to PCR product		Recombine with Donor vector containing *attP*1 and *attP*2 sites. Ensure the reading frame is maintained as shown for the triplet codons
	Sense strand (attB1)	`5'G GGG ACA AGT TTG TAC AAA AAA GCA GGC` `T-target sequence 3'`	
	Antisense strand (attB2)	`5'GGG GAC CAC TTT GTA CAA GAA AGC TGG` `GTA-target sequence`	
Gateway cloning	Directional TOPO clone	See above	Clone into TOPO activated Donor vector so that *attL*1 and *attL*2 sites now flank the insert. Ensure correct reading frame is maintained

end ligate the PCR products (*Figure 6.1*) to produce concatamers. For *Taq* DNA polymerase-generated products this will require a polishing step to ensure removal of any overhang nucleotides to create a blunt end (*Protocol 6.1*). Remember also that the PCR products must be 5'-phosphorylated for this strategy to work. Since most primers are not usually synthesized in phosphorylated format, a treatment of the primers before PCR (*Protocol 3.1*) or of the PCR product with T4 DNA kinase in the presence of ATP will be necessary for efficient self-ligation. The latter can also be performed

Table 6.2 Cleavage efficiency of some commonly used restriction endonucleases. This assay system measures the cleavage rate close to the end of duplex oligonucleotides. The restriction endonuclease cleavage site is shown in bold.

Restriction enzyme	Oligonucleotide sequence	Cleavage efficiency after 2 hour at 37°C
*Eco*RI	G**GAATTC**C	>90%
	CG**GAATTC**CG	>90%
*Xba*I	C**TCTAGA**G	0%
	GC**TCTAGA**GC	>90%
*Xho*I	C**CTCGAG**G	0%
	CC**CTCGAG**GG	10%
*Bam*HI	C**GGATCC**G	10%
	CG**GGATCC**CG	>90%

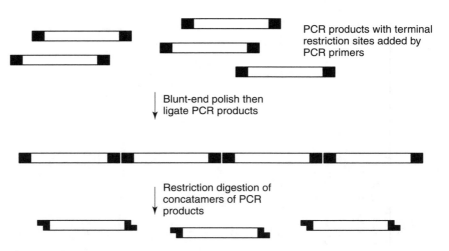

PCR products with terminal restriction sites added by PCR primers

Blunt-end polish then ligate PCR products

Restriction digestion of concatamers of PCR products

Figure 6.1

Concatemerization of blunt-end PCR products to allow efficient restriction digestion for subsequent cloning *via* cohesive ends. Restriction enzyme sites are introduced as part of the PCR primers. The PCR products are first made blunt-ended (*Protocol 6.1*) and are then ligated under conditions favoring intermolecular ligation to form concatemers. This leads to restriction sites being located within long DNA molecules, allowing efficient restriction enzyme digestion to release fragments with cohesive ends suitable for ligation into the cloning vector.

essentially according to *Protocol 3.1*, but with the primer replaced by an aliquot of PCR product. Within the resulting self-ligated concatamers, the restriction enzyme sites are now found internally within long DNA molecules and digestion of these sites is therefore more efficient. This approach can also be modified to include only half a restriction site at the 5′-end of both primers (8). If the three 3′-nucleotides of a restriction site are added to the 5′-end of each primer, the PCR products will contain half restriction sites at each end. Blunt-end ligation results in concatamers that now have full restriction sites that can be cleaved to release DNA fragments with sticky ends (*Figure 6.2*). This approach results in fragments with identical restriction sites at both ends of the digested product and therefore does not allow directional cloning.

Sometimes you will want to incorporate a restriction site within a primer for use in cloning the PCR product, but the site may occur internally within the amplified fragment. This situation may occur during some complex subcloning experiments, for example to create recombinant molecules comprising multiple components, or during cDNA cloning. In such cases it may be possible to protect the internal site(s) from digestion by incorporating 5-methyl-dCTP in place of dCTP during PCR. If the restriction enzyme is sensitive to methylation of C then only the sites in the primer, that contain unmethylated dC will be cleaved whereas the internal sites containing 5-methyl-dC will not be cleaved. The approach of protecting

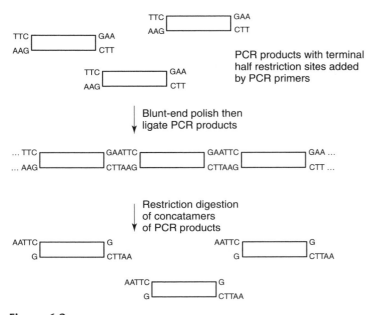

Figure 6.2

Introduction of half-restriction sites added at the 5′-ends of PCR primers, to create full sites by blunt-end ligation of PCR products. The PCR products can be made blunt-ended (*Protocol 6.1*) and then ligated to create complete restriction sites within long molecules of DNA. This allows efficient restriction enzyme cleavage to generate fragments with cohesive ends suitable for cloning into a vector.

restriction sites from restriction enzyme digestion is also used in some PCR mutagenesis strategies (Chapter 7).

The inclusion of restriction enzyme sites within primers should be carefully considered and performed *in silico* to ensure that the final construct will be as desired.

Rapid ligation

A range of suppliers produce rapid ligation components that allow 5–30 ligation reaction times for cohesive and blunt-end ligations. Examples include rapid DNA ligation kits from Roche, Fermentas and Epicentre. There is a potentially useful protocol for temperature-cycle ligation (9) in which a thermocycler is used to cycle between 10°C for 30 s and 30°C for 30 s over a period of 12–16 h. The method claims to lead to an increase of 4–8-fold in the number of colonies recovered compared with a single temperature reaction at 14°C. It seems that rather than performing the reaction for 12–16 h, a cycling reaction for 100 cycles can achieve similar increased efficiency of ligation (W. Charlton, personal communication).

TA cloning

There is a range of commercially available vectors designed for high-efficiency cloning of PCR products and many exploit selection or screening systems such as blue/white selection for recombinants. The terminal transferase activity of *Taq* DNA polymerase has been exploited for cloning purposes and the first generation of PCR cloning plasmids were designed to contain single 3′-T overhangs, enabling direct cloning of *Taq* DNA polymerase-generated PCR products which have an additional 3′-dA (4). This approach is often referred to as TA cloning (*Figure 6.3* and *Table 6.1*). Commercial systems are available from most major molecular biology reagent suppliers. A universal TA cloning method applicable for use with any vector has been described by Zhou and Gomez-Sanchez (10).

Advantages of TA cloning include: (i) no prior knowledge of the DNA sequence is necessary; (ii) post-PCR restriction digestion is not required; (iii) enzymatic blunt-end polishing of PCR products is eliminated; and (iv) the method is reliable and rapid. On the other hand, PCR fragments tend to lose their 3′-A overhang over time, even when stored at –20°C, resulting in

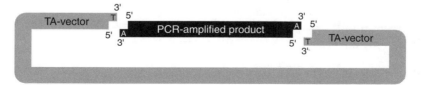

Figure 6.3

TA cloning of PCR-amplified products. The use of *Taq* DNA polymerase leads to the nontemplate-directed addition of a dA at the 3′-ends of the product. These products can be ligated into a vector with 3′-dT ends. Usually such vectors have a range of restriction sites flanking the insertion site allowing simple recloning of the fragment into another vector for further studies.

decreased cloning efficiency after prolonged storage. Also, TA cloning only works efficiently with thermostable DNA polymerases, such as *Taq* DNA polymerase, which have terminal transferase activity. However, for many PCR cloning purposes *Taq* DNA polymerase is not the most appropriate enzyme since it lacks 3′→5′ exonuclease proofreading activity and therefore can incorporate mutations into the final PCR product as discussed in Chapter 3. TA cloning does not work efficiently with many thermostable proofreading polymerases used in PCR, although some, such as Vent® and DeepVent® do yield a low proportion (~5%) of dA-tailed products.

TOPO cloning

Increased speed and efficiency of cloning dA-tailed products has also been achieved by the use of DNA topoisomerase I rather than DNA ligase. The TOPO-Cloning™ (Invitrogen) approach uses linearized vectors carrying a 3′-dT extension and preactivated with DNA topoisomerase I covalently associated with each 3′-phosphate. On addition of a PCR product the topoisomerase DNA joining activity rapidly joins the fragment into the vector allowing transformation of competent cells after only a 5 min room temperature incubation. A similar approach, Zero Blunt®, is available for blunt-ended fragments such as PCR products generated by a proofreading thermostable DNA polymerase. The problem with these approaches is that the identity between the two ends means that a fragment could be cloned in either orientation. A series of pET-based expression vectors exist for directional cloning of PCR products (Invitrogen). In this case the 5′-addition CACC is added to the upstream primer (*Table 6.1*). The downstream primer does not contain this terminal sequence. The PCR product is then incubated with TOPO-activated pET vector in which one end contains a 3′-overhang GTGG. By strand invasion this anneals to the CACC addition on the PCR product and the topoisomerase catalyses the formation of the appropriate phosphodiester link. At the other end the enzyme performs a blunt-end ligation.

Cloning long PCR fragments

The increasing use of long-range polymerase mixes (Chapter 3) is allowing the amplification of longer DNA regions that often need to be cloned for further analysis. The TOPO-XL PCR cloning kit (Invitrogen) is one option for direct cloning and relies upon the pCR®-XL-TOPO vector that carries both kanamycin and zeomycin resistance genes and the *ccd* positive selection marker. The kit also utilizes crystal violet to avoid the use of ethidium bromide and UV irradiation for detection of DNA in gels, thereby preventing DNA damage of the long molecules and enhancing isolation of full-length clones. The range of cloning kits and vectors adapted for direct cloning and expression of PCR products is increasing continually.

More rapid TOPO cloning

There are now also a series of TOPO cloning vectors that are coupled with transformation of Mach1™–T1® *E. coli* cells from Invitrogen that allow very rapid growth of a culture in just 4 h from an overnight colony. This allows DNA purification to be performed in the same day, rather than waiting for a further overnight culture.

Blunt-end cloning

PCR products can be blunt-end cloned into any vector linearized by a restriction enzyme that generates blunt ends. If a proofreading enzyme was used during PCR then the PCR products should already be blunt-ended. If *Taq* DNA polymerase was used then the dA overhangs (11) must be removed by 'polishing' the ends as described in Section 6.3 and *Protocol 6.1*.

Blunt-end cloning vectors are commercially available. For example, Perfectly Blunt™ cloning kits (Novagen) include reagents for polishing and phosphorylating the PCR product, then ligating with one of a range of linearized vectors and competent cells for transformation. The ZeroBlunt® PCR cloning kit (Invitrogen) utilizes the pCR®-Blunt vectors with a powerful selection for recombinants. The ligation of an insert into these vectors leads to an inability to express a lethal gene, *ccd*B (control of cell death). Any linearized vector molecules that are religated in the absence of an insert should be nonviable. These kits contain linearized vector, reagents for ligation and competent cells and are also available in TOPO-Cloning™ format to enhance ligation efficiency and speed. As described above, there are rapid ligation components that can be used for blunt-end ligation.

Restriction-enhanced cloning

PCR cloning vectors are available based on combining DNA ligation with restriction digestion to reduce vector only religation (*Figure 6.4*). One example is pCRScript™ plasmid (Stratagene) based on the blunt-end cloning of PCR products into an *Srf*I site. *Srf*I cuts DNA rarely and it is therefore highly unlikely that such a site will occur in the PCR product to be cloned. Including the restriction enzyme *Srf*I in the ligation reaction ensures that vector molecules remain linearized. Any vector molecules that are religated will then be redigested by the *Srf*I. However, once a PCR fragment has been ligated into the vector the *Srf*I restriction site no longer exists and the recombinant plasmid remains circularized. Subsequent transformation of *E. coli* cells with the ligation reaction results in efficient replication of the circular recombinant molecules but poor efficiency recovery of the linear vector alone. The high efficiency of this system also minimizes the ligation time, which at 1 h gives rise to a sufficient number of positive transformants for use in further analysis. Such an approach leads to recombinant plasmids in which the PCR fragment can be inserted in either orientation. An alternative strategy relies on the use of the pCRScript™ Direct vector for directional cloning of the blunt-end PCR fragment. In this case the *Srf*I-digested vector is treated with alkaline phosphatase to remove the 5'-phosphate groups. The vector is then digested with a second restriction enzyme that also generates a blunt end. This results in a vector with one dephosphorylated end and one phosphorylated end. If one of the primers is 5'-phosphorylated then directional cloning of the PCR fragment will result in ligation at both ends, resulting in a circular

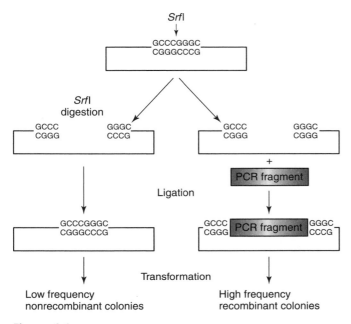

Figure 6.4

Restriction-enhanced cloning using pCRScript (Stratagene). After *Srf*I digestion the vector is linearized but in the absence of the blunt-ended PCR fragment the *Srf*I restriction site is constantly redigested with the majority of the vector staying linearized. Upon ligation of the blunt-ended PCR fragment to the vector the *Srf*I site is destroyed, generating a stable recombinant molecule that cannot be redigested by *Srf*I.

molecule with single-strand nicks. If the fragment is joined to the vector in the wrong orientation only one end can ligate but the other cannot, thus resulting in only linear molecules. Again the differential efficiency of transformation of circular versus linear molecules means high efficiency recovery of recombinants with the PCR fragment inserted in the defined orientation. The advantages of this system and other systems based on similar technologies include: (i) > 90% ligation efficiency; (ii) short ligation time; (iii) no requirement for 3′-overhangs; and (iv) the use of thermostable DNA polymerases with proofreading activity such as *Pfu* and *Pwo* DNA polymerase. *Taq* DNA polymerase can also be used, however this requires polishing the ends of the amplified DNA fragment (Section 6.3).

Ligation-independent cloning (LIC)

It is possible to generate circular recombinant molecules for efficient transformation without the need for covalent closure of the circle by DNA ligase (12–15). The simplest approach relies upon the action of terminal deoxynucleotidyltransferase (TdT) to add nontemplate directed nucleotides onto 3′-ends of DNA molecules (*Figure 6.5*). If the linearized vector is treated with TdT in the presence of, for example dGTP, then the ends of the vector will

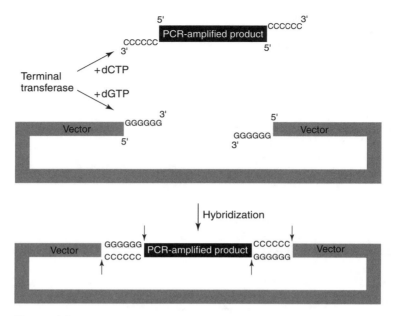

Figure 6.5

Ligation-independent cloning (LIC) of PCR fragments. Terminal transferase is used to add a single dNTP to create a polymeric tail on the 3′-ends of the vector and complementary tails on the PCR fragment. The tails can anneal leading to a stable molecule that can be directly transformed into *E. coli* cells which will repair the molecule *in vivo*.

have oligodG extensions and the vector cannot recircularize. Treatment of the PCR product with TdT in the presence of dCTP will generate oligodC extensions. The G and C tails on the vector and PCR fragment are complementary and can anneal, producing stable hybrids that can be efficiently transformed (*Table 6.1*). *In vivo* repair of the gaps will yield covalently closed molecules.

A more specific approach is to generate PCR primers that incorporate 12–15 nucleotide extensions at the 5′-ends of PCR primers designed to amplify the vector and target PCR products (*Figure 6.6*). If the extensions added to the vector and primer are designed to be complementary it is possible to subsequently join these molecules without ligation. To do this it is necessary to generate single-stranded tails corresponding to the ends of the vector and PCR products (*Table 6.1*). This is readily achieved by treating the DNA molecules with an enzyme such as T4 or T7 DNA polymerase in the presence of one dNTP determined by the design of the primers. The 3′→5′ exonuclease (proofreading) activity of the DNA polymerase removes nucleotides from the 3′-end of the DNA molecule until it encounters the first position corresponding to the dNTP added to the reaction. Once such a nucleotide is encountered the DNA polymerase may remove it, but the 5′→3′ DNA synthesis activity immediately replaces it so the DNA fragments then effectively all have the same single-stranded regions removed. The

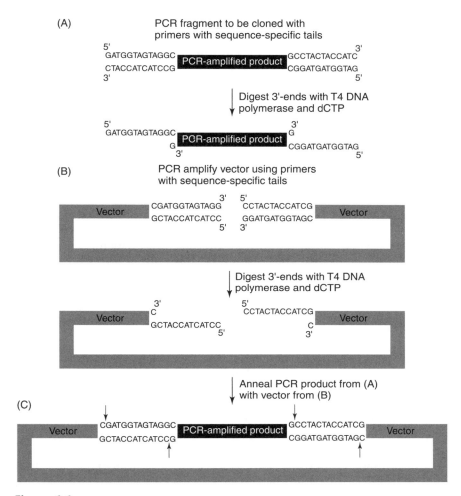

Figure 6.6

Ligation-independent cloning (LIC) of PCR fragments using specific PCR-generated tails. The primers are modified to carry 5'-additions that will become incorporated into the PCR product. The primers are designed so that treatment with T4 DNA polymerase and one nucleotide will digest the 3'-ends of the tails. The T4 polymerase will only digest the DNA until it reaches the position corresponding to the added dNTP, which allows the proofreading activity to continually repair the missing nucleotide if the polymerase removes it. This results in a defined single-stranded tail on each end of the treated fragment. (A) Preparation of the PCR product, in this case using T4 polymerase and dGTP. (B) Treatment of the amplified vector, but in this case in the presence of dCTP, yielding single-stranded tails that are complementary to those on the PCR products. (C) Annealing of the two molecules for transformation into *E. coli* cells.

single-stranded sequences on the vector and PCR products are complementary therefore allowing annealing of these ends to produce stable
double-stranded circular molecules that can be transformed efficiently.

Novagen produce an Ek/LIC vector series based on this approach and
specifically designed for expression of fusion proteins with an enterokinase
protease cleavage for recovery of the target protein. The primer sequences
are 5'-GAC GAC GAC AAG ATX-insert specific sequence 3' for the sense
strand and 5'-GAG GAG AAG CCC GGT-insert specific sequence 3' for the
antisense strand (*Table 6.1*). Position X in the sense primer must complete
the codon and therefore will encode either Met (ATG) or Ile (ATC, ATT,
ATA). Following PCR it is critical that the dNTPs are removed before
proceeding with the T4 DNA polymerase treatment. This is to ensure that
only the chosen dNTP is present at this stage of treatment to create the
single-strand ends. A PCR clean-up set is therefore necessary (Chapter 4).
The reactions with T4 DNA polymerase are performed in the presence of
dATP to create the single-strand ends that are complementary to single-
strand ends of the vector. The treated PCR fragment is then added to the
vector and annealing is allowed to proceed at 22°C for 5 min before transformation of an aliquot into *E. coli* cells. It is possible to adapt your own
preferred vector for LIC by adding appropriate sequences flanking a unique
restriction site. Restriction digestion followed by T4 DNA polymerase treatment will then yield a vector for use in LIC.

A general vector that has been shown to function efficiently in LIC for
high-throughput cloning is pMCSG7. This contains an N-terminal his-tag
followed by a tobacco etch virus (TEV) protease cleavage site and a *Ssp*I
(AAT/ATT) restriction site for LIC (16).

An alternative approach to generating the single-stranded ends is to
generate PCR primers with about 12 nucleotide 5'-tails that are synthesized
to replace dT with dU. The resulting PCR fragment, synthesized using the
four normal dNTPs, therefore has one strand carrying terminal dUs while
the complementary ends would contain dTs. Treatment of the PCR
products with the enzyme uracil *N*-glycosylase leads to the cleavage of
N-glycosidic bonds between uracil and deoxyribose leading to a loss of the
base. The loss of bases leads to disruption of base pairing and therefore
exposure of single strands that can be annealed to a vector displaying
complementary single-stranded ends. It is possible to clone either non-
directionally by having identical ends on both primers, or directionally by
having distinct sequences on the two primers used in the original PCR, and
by appropriate choice of a directional vector.

These approaches to ligation-independent cloning can be performed
more rapidly than many traditional ligation reactions as the annealing of
single-stranded ends is achieved within about 30 min to produce trans-
formation-ready DNA molecules.

Recombination-mediated cloning

This approach is intended to reduce the need to use traditional restriction
digestion and ligation steps in gene cloning procedures and to support high-
throughput cloning applications based on robotic platforms. The most
prominent system is the GATEWAY™ system from Life Technologies, now

part of Invitrogen. There is a useful online tutorial that explains the technology at http://www.invitrogen.com/Content/Online%20Seminars/gateway/gatewayhome.html. This system provides a generic and rapid mechanism for initial cloning of PCR fragments into an entry clone by utilizing a recombinase and negative selection. The insert can subsequently be transferred to a wide range of expression clones by recombination from the entry clone. This allows the parallel production of a range of expression clones that can be tested in a range of different expression systems from bacteria to mammalian cells. The approach is rapid and is adaptable for high-throughput cloning. The simplicity of the Gateway approach relies upon *att* recombination sites that are built into PCR primers and the vectors (*Table 6.1*) and that are acted upon by enzyme mixes that cleavage and rejoin *att* sites.

The system is based upon the naturally occurring bacteriophage lambda system for integration of its phage genome into the *E. coli* chromosome to enter the lysogenic phase and subsequently its excision from the genome to enter the lytic cycle. These integration and excision events occur by homologous recombination controlled by enzymes. The recombination sites in lambda are known as *attP* and those in *E. coli* are *attB* and are only 25 bp long. The integration reaction involves two proteins, an integrase and a host integration factor. Homologous recombination leads to the production of *att*L and *att*R sites with no loss of DNA sequence. These sites contain a core sequence of 15 nucleotides that is the recognition sequence for the integrase. The excision process involves the action of the two enzymes above plus an excisionase. There has been significant refinement of the *att* sites so that a range of compatible sites now exist. This provides specificity and directionality to cloning reactions.

There are two processes for creating an entry clone from PCR products:

- TOPO cloning;
- adding *att* sites to the PCR primers.

TOPO cloning approaches have been discussed earlier in this Section and a range of TOPO-ready Gateway vectors are available for directional cloning. The consequence of cloning a PCR fragment in this manner is to place the DNA adjacent to appropriate *att*L1 and *att*L2 sites that are present in the vector. The critical consideration is to ensure that the resulting clone will be in-frame with the sequence upstream in the final vector if there is reliance on a vector-based ribosome binding site and ATG. *In silico* cloning experiments using Vector NTI allow the various constructs to be designed to ensure they will be appropriate before ordering the PCR primers.

The other approach to creating an entry clone involves adding *att*B sites to the 5′-ends of the PCR primers. The *att*B sequences are slightly different and are represented as *att*B1 and *att*B2 that recombine with the *att*P1 and *att*P2 sites respectively on the donor vector, creating *att*L1 and *att*L2 sites when BP clonase is added (*Figure 6.7*). This is called the BP reaction.

The subsequent movement of the insert from the entry clone, in which it is transcriptionally silent, into appropriate expression vectors is facilitated by LR clonase. This allows the movement of the insert from the entry clone into various expression vectors that carry *att*R1 and *att*R2 sites with which the *att*L1 and *att*L2 sites of the entry clone can recombine to recreate *att*B1

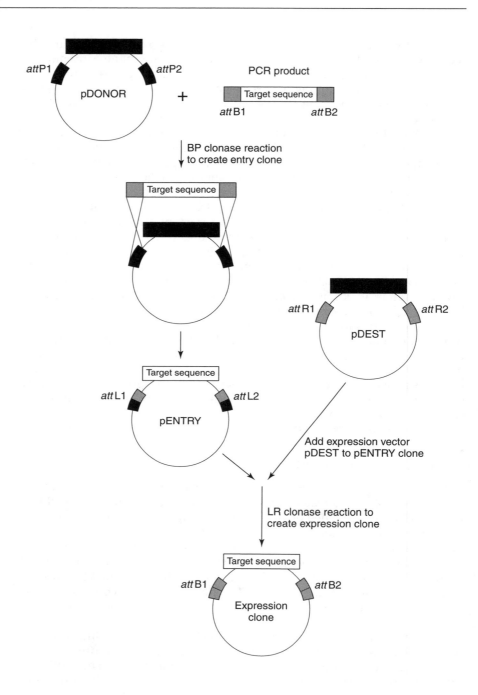

and *att*B2 sites in the expression clone (*Figure 6.7*). This is called the LR reaction.

There is a powerful positive selection system associated with the various cloning steps. The vectors contain the *ccd* gene between its *att* sites, which confers a lethal phenotype when the vector is transformed into a normal *E. coli* strain. This means that only clones that have undergone recombination and which have therefore lost the *ccd* gene and at the same time gained the target insert, should be viable following transformation.

There are a wide a range of expression vectors available for final expression in bacteria, yeast or various animal cell lines. In addition it is possible to adapt your own vector to accept inserts by Gateway cloning. A series of cassettes are available that can be cloned into your favorite vector to create a vector in the correct reading frame for recombinational insertion of a PCR fragment or pENTRY clone.

Primers for the gene-specific regions should be designed according to the normal rules for primer design (Chapter 3), and so would normally be around 15–20 nt long. The *att* sites are added as additional sequence to the 5'-end. If primers are ordered from Invitrogen then the appropriate *att* sequences can be selected automatically during the ordering process and the cost of synthesis is relatively cheap.

An example of appropriate design of primers to add *att*B1 and B2 sites to the 5' and 3' ends respectively of a gene of interest are shown in *Table 6.2*. The addition of *att* sites leads to the inclusion of additional amino acids at the start and end of the encoded protein. If stop codons are introduced before the 3'-*att*B2 site then it is possible to ensure that the C-terminal end of the protein finishes at the natural sequence.

6.5 Confirmation of cloned PCR fragments

Assume that you have cloned a PCR fragment and have analyzed the resulting clones by PCR screening of bacterial colonies or cultures (*Protocol 6.2*), or by preparing plasmid DNA and releasing the insert by restriction digestion, and you identify clones with the correct fragment size. A word

Figure 6.7 (opposite)

Gateway cloning. A PCR product is generated with primers containing 5'-terminal *att*B sites. This PCR product is purified and mixed with a pDONOR vector that carries *att*P sites. Addition of BP clonase leads to homologous recombination of the PCR fragment into the pDONOR plasmid with the creation of *aat*L sites. There is also excision of the corresponding vector fragment that carries a lethal *ccd*B gene. Any pDONOR vector that has not undergone recombination leads to death of any bacterial transformant. The resulting pENTRY clone carrying the PCR insert is isolated from bacterial transformants, purified and then mixed with an appropriate expression vector often called pDEST vectors. These pDEST vectors also carry a *ccd*B insert flanked by *att*R sites. Addition of LR clonase leads to homologous recombination between the *att* sites. The insert is inserted into the pDEST vector to form an expression clone. Once again any pDEST vector that fails to recombine will be lost during transformation due to the *ccd*B gene. Careful design of the original PCR primers should lead to a clone that is in the correct reading frame for efficient expression in the chosen host.

of caution: for the reasons described in Chapter 5 do not simply rely on restriction digest analysis by using the restriction enzymes used to clone your amplified product. All this will tell you is that you have the correct sized insert, but not whether it is the correct fragment. Perhaps the PCR amplification resulted in a heterogeneous mixture of DNA fragments with near-identical sizes that you have cut out from the gel, purified and now cloned into a suitable vector. If you engineered restriction sites into your primers then any DNA fragment, including nonspecific products, will also contain these sites. A similar result would be obtained if you blunt-end cloned your PCR fragment and analyzed it using restriction enzymes that cut within the polylinker of the plasmid. It is important to perform your PCR reaction under high stringency conditions to minimize the possibility of generating other DNA fragments. It is essential to analyze your clones using known internal restriction enzymes or nested PCR. Several clones should be analyzed to verify that the amplified band used for the cloning experiment consisted of one DNA species. For putative positive clones you should also then determine the DNA sequence of the insert to ensure that the DNA fragment is exactly what you expect.

References

1. Sambrook J, Fritsch EF, Maniatis T (1981) *Molecular Cloning: A Laboratory Manual* (2nd edition). Cold Spring Harbor Laboratory Press, Cold Spring Harbor, NY.
2. Vogelstein B, Gillespie D (1979) Preparative and analytical purification of DNA from agarose. *Proc Natl Acad Sci USA* **76**: 615–619.
3. Yang RCA, Lis J, Wu B (1979) Elution of DNA from agarose gels after electrophoresis. *Methods Enzymol* **68**: 176–182.
4. Marchuk D, Drumm M, Saulino A, Collins FS (1991) Construction of T-vectors, a rapid and general system for direct cloning of unmodified PCR products. *Nucleic Acids Res* **19**: 1154.
5. Pan G, Luetke K, Juby CD, Brousseau R, Sadowski P (1993) Ligation of synthetic activated DNA substrates by site-specific recombinases and topoisomerase I. *J Biol Chem* **268**: 3683–3689.
6. Scharf SJ, Horn GT, Erlich HA (1986) Direct cloning and sequence analysis of enzymatically amplified genomic sequences. *Science* **233**: 1076–1078.
7. Moreira RF, Noren CJ (1995) Minimum duplex requirements for restriction enzyme cleavage near the termini of linear DNA fragments. *BioTechniques* **19**: 56–59.
8. Kaufman DL, Evans GA (1990) Restriction endonuclease cleavage at the termini of PCR products. *BioTechniques* **9**: 304.
9. Anders H, Lund AH, Duch M, Pedersen FS (1996) Increased cloning efficiency by temperature-cycle ligation. *Nucleic Acids Res* **24**: 800–801.
10. Zhou M-Y, Gomez-Sanchez CE (2000) Universal TA Cloning. *Curr Issues Mol Biol* **2**: 1–7 http://www.horizonpress.com/cimb/abstracts/v2/01.html
11. Clark JM (1988) Novel non-templated nucleotide addition reactions catalyzed by prokaryotic and eukaryotic DNA polymerases. *Nucleic Acids Res* **16**: 9677–9686.
12. Aslanidis C, de Jong PJ (1990) Ligation-independent cloning of PCR products (LIC-PCR). *Nucleic Acids Res* **18**: 6069–6074.
13. Aslanidis C, de Jong PJ, Schmitz G (1994) Minimal length requirement of the single-stranded tails for ligation-independent cloning (LIC) of PCR products. *PCR Methods Appl* **4**: 172–177.

14. Haun RS, Moss J (1992) Ligation-independent cloning of glutathione S-transferase fusion genes for expression in *Escherichia coli. Gene* **112**: 37–43.
15. Haun RS, Serventi IM, Moss J (1992) Rapid, reliable ligation-independent cloning of PCR products using modified plasmid vectors. *BioTechniques* **13**: 515–518.
16. Stols L, Gu M, Dieckman L, Raffen R, Collart FR, Donnelly MI (2002) A new vector for high-throughput, ligation-independent cloning encoding a tobacco etch virus protease cleavage site. *Protein Expr Purif* **25**: 8–15.

Protocol 6.1 Blunt-end polishing of PCR fragments

MATERIALS

T7 DNA polymerase

10 × T7 DNA polymerase buffer

100 mM $MgCl_2$

2 mM each dNTP mix

Distilled water

EQUIPMENT

Microcentrifuge tube

Microcentrifuge

Water bath or heat block set at 70°C

1. Combine in a microcentrifuge tube:
   - up to 40 µl of purified PCR product;
   - 5 µl of 10 × T7 polymerase buffer;
   - 2 µl of 100 mM $MgCl_2$;
   - 2.5 µl of 2 mM dNTP mix;
   - 1 unit of T7 DNA polymerase;
   - water to 50 µl.

2. Collect reagents and mix by a brief (1 s) centrifugation step.

3. Incubate at room temperature for 30 min.

4. Heat to 70°C for 10 min to inactivate the polymerase.

Protocol 6.2 PCR screening of bacterial colonies or cultures

MATERIALS

Agar plate with colonies for screening

Sterile water

PCR reactant mix containing 5 pmol each primer, 0.2 mM each dNTP, 1 × PCR buffer, 0.5 unit *Taq* DNA polymerase, water per 25 μl[1]

LB or 2 × TY media with appropriate antibiotic used to grow cultures

Agarose gel

EQUIPMENT

0.5 or 0.2 ml tubes or microtitre plates for PCR

Microcentrifuge

Orbital incubator for growth of bacterial cultures

It is possible to directly analyze colonies to determine whether they contain a desired recombinant plasmid and at the same time to set up overnight cultures from the colonies to allow isolation of recombinant plasmids for further study.

1. Prepare 3 ml aliquots of LB or 2 × TY media containing the appropriate antibiotic in 20–50 ml volume plastic tubes.

2. Pipette 100 μl of sterile water into a 0.5 ml microcentrifuge tube.

3. Use a toothpick to collect a colony from the agar plate and swirl gently in the sterile water, then use the toothpick to inoculate a tube containing growth medium[2].

4. Place the tube containing the water and dispersed bacterial suspension in a thermal cycler and heat to 100°C for 5 min then cool to 20°C.

5. Centrifuge for 1 min at 13 000 *g* in a microcentrifuge to pellet the debris.

6. Pipette a 24 μl aliquot of the PCR premix into a PCR tube and add 1 μl of the DNA solution from the bacterial colony.

7. Collect and mix reagents by a brief (1 s) centrifugation step.

8. Perform the PCR using the following conditions:
   - 94°C, 5 min initial denaturation;

- 94°C, 30 s[3]; 55°C[4], 30 s[3]; 72°C, 30 s[3] for 25–30 cycles;
- 72°C, 1 min.

9. Analyze a 5 µl aliquot on an agarose gel.

10. Meanwhile place the inoculated cultures in an orbital incubator at 37°C to allow growth of the culture for subsequent plasmid isolation from PCR positive colonies, or incubate the agar plate to allow subsequent culture inoculation once positives have been identified.

NOTES

1. A hot-start procedure (Chapter 4) will enhance specificity, but the short cycle times often used make this approach unsuitable for TaqStart antibody approaches.

2. It is advisable to use the water lysis approach, but it is also possible to add a small amount of a bacterial pellet directly to a PCR tube. Touch a toothpick to the edge of the colony taking care not to transfer a visible amount of the colony. Swirl the toothpick briefly in the PCR mix and then if required inoculate a tube containing growth medium such as 2 × TY or LB, to allow growth of a culture, or streak onto an appropriate agar plate.

3. The length of each incubation step depends upon the type of thermal cycler. In some cases for reaction temperature monitoring instruments the time may be as short as 1–10 s.

4. The annealing temperature depends upon the length of the primers and should be as high as possible to ensure specificity of priming. It is worth testing a positive culture at different annealing temperatures. This can conveniently be undertaken with a gradient thermal cycler.

PCR mutagenesis

7

7.1 Introduction

The ability to mutate DNA by changing its nucleotide sequence is an important molecular biology tool that has led to significant insights into:

- the structure–function relationships of protein, DNA and RNA molecules;
- molecular interactions involving biomolecules including receptor–ligand, antibody–antigen and enzyme–substrate, effector or regulator complexes;
- *in vivo* regulation of gene expression.

In addition, it provides a routine approach for the modification of DNA sequences to facilitate cloning or analysis to make life simpler for the scientist.

The structural and functional importance of specific nucleotide sequences in a DNA or RNA molecule, or of a specific amino acid within a protein, can be explored by changing the corresponding DNA sequence by site-directed mutagenesis. This requires some knowledge of positions that are worth investigating. In the case of proteins it is usually important to have some structural information. However, where there is limited or no structural information then random mutagenesis techniques can be employed to explore biomolecule function. This involves the random changing of nucleotides and screening for those that show interesting differences based on some phenotype, such as an enzyme or binding assay. Interesting variants can then be sequenced to see what nucleotide change(s) has led to the new phenotype. This approach is very powerful for generating new proteins for biotechnology applications.

PCR mutagenesis allows us to modify and engineer any target DNA with ease and high efficiency and can be used to:

- introduce a deletion or insertion of sequence information including restructuring genes by domain swapping experiments;
- alter one or a few specific nucleotides; or
- randomly mutate a region of nucleotide sequence including a complete coding region if required.

There are a range of kits available for mutagenesis experiments. It can be convenient to use a kit when you require a mutation as a one-off experiment, or when funds are not limiting. Sometimes when you must undertake extensive mutagenesis reactions the cost of purchasing kits can be prohibitive. Normally, the kits are based on published procedures and therefore it is possible to adapt the approach and build your own 'kits' using individual reagents that you can optimize for your studies.

For any mutagenesis reaction the template DNA must be single-stranded to allow an oligonucleotide to bind to its complementary sequence so that it can act as a DNA synthesis primer. Historically this required cloning your gene of interest into a vector such as bacteriophage M13, which produces single-stranded copies of its genome, and any inserted gene, which can be isolated from the culture medium. Alternately double-stranded plasmid DNA can be denatured by using an alkali treatment and a neutralization step or by heating. PCR is ideally suited to mutagenesis because the production of single-strand template occurs during the denaturing step of the reaction. Even though most protocols now use a proofreading thermostable DNA polymerase, it is still appropriate to limit the number of cycles of amplification to limit possible errors becoming amplified. This can be done quite simply by performing the reaction on a relatively high concentration of the recombinant plasmid template (around 10–100 ng).

The amplification of DNA in a PCR means that at the end of the reaction the newly synthesized, mutated product should be present in excess over the wild-type template. However, most mutagenesis approaches incorporate a step that acts to destroy the original template DNA, leaving the mutated DNA to be transformed into the host cells. This is important for ensuring that most clones recovered are derived from the mutated DNA rather than the wild-type template. As we will see some mutagenesis procedures, such as the Quikchange approach, use a PCR-type reaction but lead to only linear rather than exponential amplification of the product. Selecting against the wild-type template is therefore critical in these experiments.

In this Chapter we shall cover the basic principles of PCR-based mutagenesis for introducing a variety of DNA modifications.

7.2   Inverse PCR mutagenesis

The introduction of mutations by inverse PCR involves synthesis of the whole plasmid. This allows a plasmid to be altered by including a mutation during the PCR replication of the complete plasmid followed by ligation to reform the circular plasmid molecule as outlined in *Figure 7.1* and *Protocol 7.1* (1). In essence the two primers are designed to anneal to opposite strands of the plasmid so that their 5'-ends are adjacent. These primers will thus lead to copying of the two strands of the plasmid. The 5'-ends of both primers must be phosphorylated by T4 DNA kinase treatment (*Protocol 3.1*) so that they can be ligated to the 3'-end of the replicated plasmid strand. The original procedures used *Taq* DNA polymerase but this can lead to a nontemplate directed addition of a 3'-dA that could introduce an unselected secondary insertion mutation. This additional dA can be removed by treatment of the DNA with Klenow fragment of DNA polymerase I in the presence of the four nucleotides. However, to avoid this problem and for higher fidelity a proofreading DNA polymerase (Chapter 3) should be used. It is important to ensure sufficient time for the copying of the template given the length of the plasmid template and typically one should allow 2 min per kb. This method relies upon circularization of the PCR products by blunt-end ligation.

Copying a whole plasmid by inverse PCR can be used to introduce point mutations, deletions or insertions. Since the wild-type plasmid is used as template it is usual to start with a low concentration of template so that it

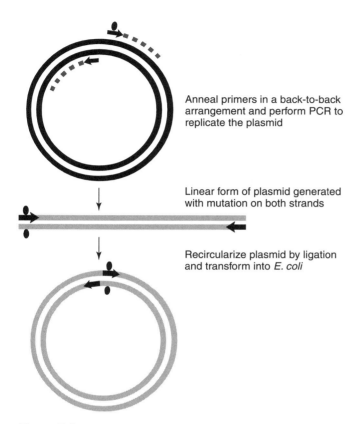

Anneal primers in a back-to-back arrangement and perform PCR to replicate the plasmid

Linear form of plasmid generated with mutation on both strands

Recircularize plasmid by ligation and transform into *E. coli*

Figure 7.1

Inverse PCR for mutagenesis of a plasmid. Two primers are designed in a back-to-back manner (with their 5′-ends designed to correspond to adjacent positions). One primer carries a point mutation (●). PCR amplification of the complete plasmid using a proofreading DNA polymerase will result in linear double-stranded, mutated plasmid molecules that can be blunt-end self-ligated to give circular molecules for transformation into *E. coli*. Alternatively if a deletion is to be introduced then the primers are designed to flank the sequence to be deleted.

will not contribute a high background during screening. For example with a plasmid miniprep the DNA should be diluted 1:1 000 or 1:10 000 and 1–10 µl used in the reaction. However, it is more efficient to start with a higher concentration of plasmid template and perform fewer PCR cycles. This approach can be used if it is coupled to a procedure to destroy the wild-type DNA once its function as a template has been performed. There are several such 'high efficiency' methods that have been developed to select against the template. The one that has become most widely adopted and is of general applicability is based on the Quikchange procedure developed by Stratagene. This approach is outlined in detail below, and could be adopted to enhance the efficiency of inverse PCR by including a *Dpn*I digestion step after the ligation step.

Quikchange mutagenesis

The Quikchange strategy makes use of an inverse site-directed mutagenesis approach together with destruction of parental template by using methylation-dependent restriction endonuclease (*DpnI*) digestion (*Figure 7.2* and *Protocol 7.2*). The principle can be adapted for other plasmid systems and reagents. Although the process is performed in a thermocycler, using typical reagents for a PCR, due to the complementary nature of the primers used it

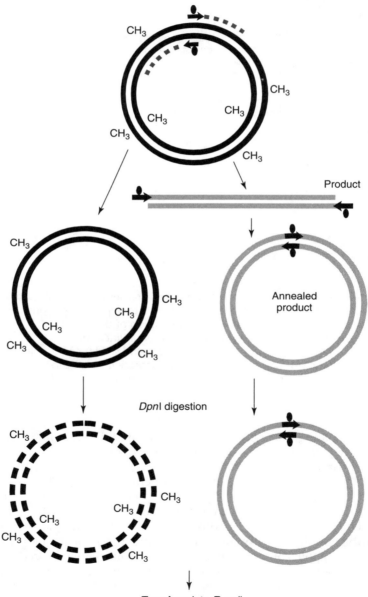

is actually a linear amplification rather than an exponential amplification process. This is because the primers are unable to productively anneal to the products of the reaction. Nonetheless, it is a highly powerful strategy that works efficiently and one which you could adapt to many PCR cloning and mutagenesis applications. The approach requires two primers (25–45 nt) that are perfectly complementary and with a $T_m \geqslant 78°C$. So when introducing nucleotide changes it is necessary only to design the sequence of one primer, which includes the required mutations. The other primer is simply the reverse complement of this designed sequence. The changes will therefore be introduced into both strands of the newly synthesized plasmid DNA. One primer acts as the priming site for synthesis of one strand whilst the other primer acts as the priming site for the complementary strand. The resulting products are linear single strands that can anneal to form full-length plasmids with single-stranded overhangs corresponding to the primer sequences. These molecules can therefore circularize by base pairing of the complementary primer sequences which carry the desired mutations. The length of the single-strand tails, 25–45 nt, results in a stable duplex and so no ligation step is necessary and thus the primers do not need to be phosphorylated before use.

The reaction is treated with the restriction endonuclease *Dpn*1 which recognizes the 4 base-pair sequence GATC, but will only cleave its recognition sequence if the adenine residue is methylated. Template plasmid DNA that was isolated from a *dam*⁺ strain of *E. coli* will therefore be a substrate for *Dpn*1, but the *in vitro* synthesized 'mutant' DNA will not be methylated and so will not be cleaved. This means the wild-type template DNA is extensively digested (a 4 bp recognition enzyme such as *Dpn*I will cut on average every 256 bp) while the *in vitro* amplified nonmethylated mutant DNA remains intact. Of course it is therefore essential to isolate the original plasmid DNA from a *dam*⁺ strain. Transformation of an aliquot of the *Dpn*I digest reaction into *E. coli* cells leads to a high efficiency of mutant plasmid isolation (usually >80%). As the template DNA will be destroyed, a high concentration of starting plasmid template DNA (up to 100 ng) can be employed to minimize the number of DNA synthesis cycles. The Quikchange approach can be used for introducing point mutations, insertions or deletions by design of appropriate primers (Sections 7.5–7.7). A protocol for this method is provided in *Protocol 7.2*. This

Figure 7.2 (opposite)

Quikchange PCR-based site-directed mutagenesis approach from Stratagene. The whole plasmid is copied to introduce a mutation by the use of complementary mutagenic primers. The template DNA is a plasmid isolated from a *dam*⁺ *E. coli* strain and so is methylated. Because the primers are complementary to one another the reaction is actually a linear amplification rather than the normal exponential amplification of a PCR. Only the original template DNA can be copied at each cycle as the products of the previous reaction cannot act as templates. The resulting plasmid length fragments have overhanging single-strand ends that correspond to the mutagenic primers, and these can anneal to form open circular molecules. The DNA is digested with the restriction enzyme *Dpn*I that will cleave its restriction sites only if the strand is methylated, but will not cleave methylated DNA. This results in wild-type molecules being digested but the newly synthesized DNA which is not methylated is protected from digestion. The reaction products are transformed into *E. coli* and this procedure leads to a high efficiency of recovery of mutant clones.

uses KOD DNA polymerase (Novagen) rather than the *Pfu*Turbo recommended in the kit, since in our experience KOD provides a much more robust and efficient DNA polymerase with high yields of product.

General applicability of *Dpn*I selection

The ability to destroy the template DNA by using *Dpn*I in this way could be built into many PCR methodologies and so we recommend that when

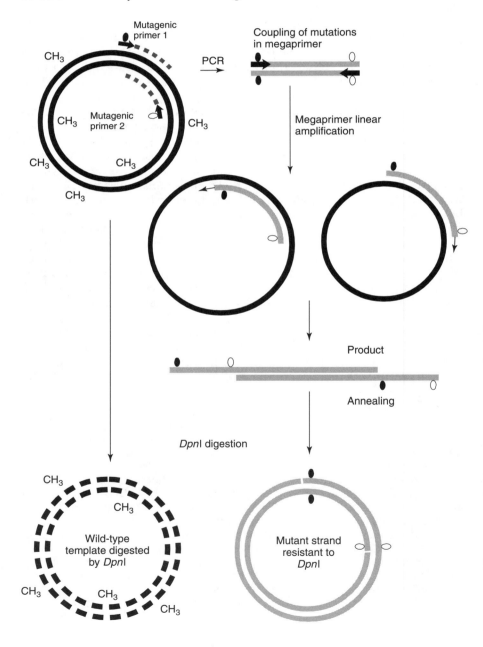

considering strategies for new experimental procedures you should consider whether a template destruction step such as this would be helpful. It may for example allow you to cut out a purification step which will prevent some level of product loss.

Coupling independent mutations

In some cases you may wish to introduce more than one mutation into a target gene but the mutation sites are too far apart to be included in the same mutagenic primer. A modified Quikchange approach can be used. As outlined in *Figure 7.3*, two primers are designed, one to incorporate each of the desired mutations into the target gene. These primers are designed to point towards each other to facilitate the amplification of the intervening DNA. This product now has both mutations physically linked in a single DNA fragment. The product can be purified by gel extraction or a PCR clean-up kit. It is then annealed to the original plasmid DNA and both strands act as primers for synthesis of the remainder of the plasmid molecule. Digestion of the reaction mix with *Dpn*I leads to destruction of the wild-type plasmid molecules. The products of the reaction can anneal to form double-stranded molecules with single-strand tails corresponding to the original PCR product which can therefore anneal to circularize the plasmid. This digested DNA mixture is then used to transform *E. coli* cells to yield mutant DNA with high efficiency.

An alternative strategy is to perform one round of mutagenesis with one set of mutagenic primers both of which are 5'-phosophorylated. After *Dpn*I digestion, the reaction is treated with DNA ligase and *dam* methylase in the presence of S-adenosylmethionine, to create closed circular DNA that is then methylated. Following a clean-up step this acts as template for the next set of mutagenesis primers (2). The approach allows sequential introduction of mutations into the template DNA without intermediate transformation and plasmid characterization steps.

Stratagene have also introduced a Multisites kit, which uses a series of phosphorylated primers in a Quikchange reaction containing a thermostable ligase. Only one primer is designed per mutation, and only one strand of the template is copied. DNA extension from one primer proceeds until it reaches the adjacent primer annealed to the template, which has also been extended. The ligase can then join these DNA fragments. DNA synthesis from the 3'-most primer continues until it proceeds around the rest of the plasmid template and reaches the 5'-most primer. The wild-type template is destroyed by the *Dpn*I digestion before transformation.

Figure 7.3 (opposite)

Introduction of coupled mutations by use of a megaprimer approach and Quikchange selection. Two mutagenic primers are designed to amplify a region of DNA, each primer introducing a mutation (●) and (○). This fragment is then used as a megaprimer to amplify the remainder of the plasmid in a manner analogous to the Quikchange approach. In this case the primers are longer DNA fragments rather than short oligonucleotide primers. Subsequent restriction digestion with *Dpn*I leads to destruction of the wild-type, methylated template DNA, but not the newly synthesized plasmids that circularize to form open circles due to the complementarity of the megaprimer sequences.

7.3 Unique sites elimination

This approach relies upon the coupled removal of a unique restriction site together with the introduction of the desired mutation. It can be adapted for essentially any plasmid that carries a unique restriction site. If the restriction enzyme site lies within a coding region, then it is important to ensure that the change you will introduce to destroy the site does not alter the coding sequence of the DNA, otherwise you will introduce another mutation into the encoded protein. As outlined in *Figure 7.4*, two primers are designed, one to incorporate the desired mutation and the other to eliminate the unique restriction site. These primers are designed to point towards each other to amplify the intervening DNA. This product now has both mutations physically linked on a single DNA molecule. The product can be purified by gel purification or a PCR clean-up kit. It is then annealed to the original plasmid and acts as a primer for synthesis of the remainder of the plasmid molecule. This is a linear amplification reaction of essentially the same type as the Quikchange approach (Section 7.2). The resulting fragments will therefore have long single-strand ends that can anneal to form circular molecules. There is no need to perform a ligation step. If you do wish to ligate, then ensure that the primers are 5'-phosphorylated before performing the PCR. Next digest the DNA with the restriction enzyme whose site you have destroyed. This will linearize any wild-type plasmid molecules, but any in which the site has been destroyed will not be linearized. If you have performed a ligation, ensure that you have performed a heat inactivation step before the digest to prevent any religation of wild-type sequences. This digested DNA mixture is then used to transform *E. coli* cells. Since circular DNA is able to transform much more efficiently (~3 000-fold greater) than linear DNA molecules, the cells will largely be transformed by the circular mutant molecules. Individual colonies can then be screened for plasmid that has lost the restriction site and should therefore contain the linked desired mutation. An alternative approach is to grow a culture of the transformation cells without plating out. Plasmid DNA can then be isolated from these cells and subjected to a restriction digest with the enzyme whose site was knocked out. Any wild-type plasmid that was able to replicate will again be linearized, but mutant DNA will remain circular. A further transformation step with this DNA will effectively ensure that all colonies will now contain the mutation.

7.4 Splicing by overlap extension (SOEing)

PCR SOEing can be used to join or 'splice' together sequences, such as a regulatory sequence with a coding region, or two protein domain coding regions. It can also introduce deletions, insertions or point mutations into a DNA sequence (3). *Figure 7.5* shows a schematic example of the approach for joining two sequences together. The critical aspect is that at least one primer (P3 in *Figure 7.5*) should contain a sequence overlap with the end of the fragment to which it is to be joined.

Two separate PCRs are performed, often simultaneously, using the appropriate template DNAs and appropriate primers; one reaction with primers

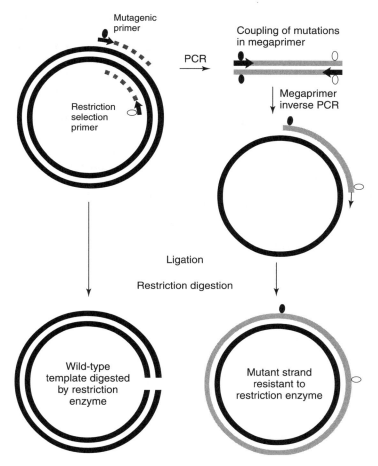

Figure 7.4

Unique site elimination (USE) mutagenesis scheme using PCR. Two primers are used to introduce the desired mutation (●) and the selection mutation (○) that will destroy a unique restriction site. This PCR step physically joins these two mutations on a single DNA molecule that can be purified and used as a primer for synthesis on a plasmid template. The resulting heteroduplex molecules are restriction digested with the enzyme whose site has been mutated. Wild-type molecules that contain an intact restriction site will be linearized but heteroduplex molecules carrying one wild-type and one mutant strand will remain circular and will transform *E. coli* cells efficiently. Plasmid replication in *E. coli* will resolve the strands leading to some cells carrying wild-type plasmids and some carrying mutants. The cells are pooled, and plasmid DNA is purified and digested with the selection enzyme. The wild-type plasmids will be linearized, but the mutant plasmids will remain circular and can be recovered as clones following a second round of *E. coli* transformation.

P1 and P2 and a second with primers P3 and P4. The design of at least one, and sometimes two 'overlap' primers (P2 and P3) leads to two PCR products which have a region of identical sequence, defined by the overlap primers. Aliquots of these two primary reactions are mixed in a second PCR. In the early rounds of this PCR the products of the primary PCRs are denatured and can anneal with each other, due to the overlap, to create 3'-ends that can be extended on the complementary template strand to give a full-length product. Once this product is formed the flanking primers (P1 and P4; *Figure 7.5*) will allow exponential amplification of this full-length product.

As shown in *Figure 7.5* only one strand of each PCR product is 'productive'. These are the strands that have 3'-ends capable of priming on the complementary strand of the other product. The other two strands cannot act in this way and so are considered to be nonproductive. However, primers P1 and P4 will be able to anneal to these nonproductive strands, which can act as templates for linear amplification of further copies of the productive

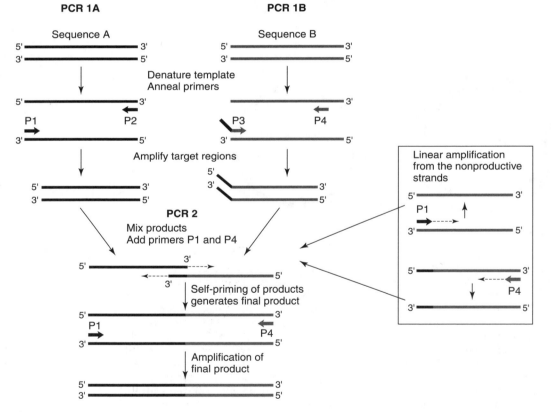

Figure 7.5

Example of gene SOEing experiments to join two sequences A and B. Initial PCR amplification of gene A uses two primers P1 and P2, and amplification of gene B uses two primers P3 and P4. Note that primer P3 contains a 5'-extension corresponding to the sequence on gene A at the site of joining. The two fragments can therefore anneal by virtue of the complementary terminal tails and can self-prime with subsequent amplification of the full-length product by flanking primers P1 and P4. As shown in the box, only one strand of each product is 'productive' for self-priming.

strands as shown in the box in *Figure 7.5*, thus enhancing the likelihood of formation of a full-length product.

This basic approach to PCR SOEing can also be used to introduce mutations into a DNA sequence. In this case the template DNA for both primary PCRs will be the same. The overlap primers (P2 and P3) will be designed to incorporate the mutations and so the overlaps will be complementary but both strands will now include the appropriate base changes (*Figures 7.6* and *7.7*). In the secondary PCR the mutated products of the primary PCRs will anneal with one another and self-prime to form full-

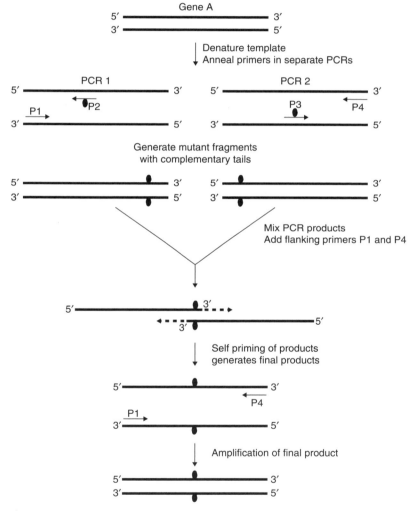

Figure 7.6

Example of gene SOEing experiment to introduce a mutation (●). Parallel PCRs introduce the mutation into two products by use of primers P1 and P2, and P3 and P4. Primers P2 and P3 are complementary and therefore facilitate self-priming of the PCR products. The full-length mutated product is amplified by flanking primers P1 and P4.

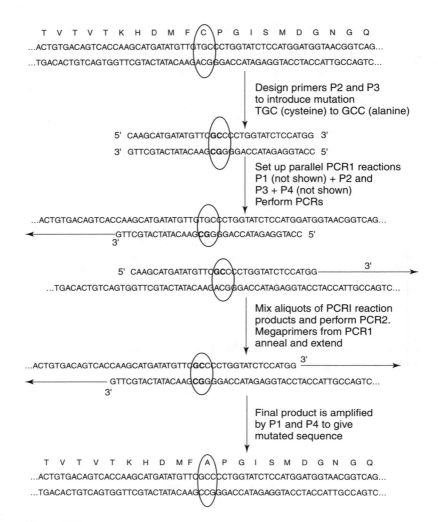

Figure 7.7

Example of changing a DNA sequence by gene SOEing focusing on the region to be mutated. A region of DNA sequence is shown and a position to be mutated is identified. Two primers, P2 and P3 are designed to be complementary and to introduce the required nucleotide changes. Primer P2 is used with P1 (not shown) in a PCR, and primer P3 is used with P4 (not shown). Both PCRs use the same template DNA and therefore generate two parts of the gene, both containing the desired mutation, and containing a region of overlap defined by primers P2 and P3. The two products are mixed and used in a further PCR in which the productive strands anneal and self-prime. Further amplification of the full-length product is achieved by primers P1 and P4. The resulting DNA product carries the desired mutation.

length, mutated product, which can be amplified by the flanking primers (P1 and P4). This product can then be digested with appropriate restriction enzymes and inserted in place of the corresponding wild-type region in the starting plasmid. Alternatively it could be introduced by an adaptation of

Quikchange procedure with the fragment acting as megaprimers for plasmid synthesis, followed by *DpnI* digestion and transformation (Section 7.2).

Primer design

The most important parameter for successful SOEing is primer design. The same general rules as described in Chapter 3 are applicable, but there are a number of further considerations. For the primary PCRs, the flanking primers (P1 and P4; *Figures 7.5* and *7.6*) should be designed as 'normal' primers perhaps 18–20 nucleotides in length and these could be vector-specific primers if the whole insert is to be amplified. Both of the internal primers (P2 and P3) should contain the mutation(s) (*Figure 7.6*) or contain a 5'-addition that is complementary to part of the DNA fragment to which it is to be joined (*Figure 7.5*). Since P2 and P3 will act as the primers during the secondary PCR, they should be around 25–35 nt long depending upon the number of mutations being introduced, or if they are being used to join fragments then the added tails should be around 15–20 nt long. Essentially primers P2 and P3 can be complementary if they are acting on the same template and are introducing mutations. If they are being used to join fragments then the 3'-regions should be sequence specific, but the tails should be complementary to one another, probably representing a region of one of the sequences being joined. A basic procedure for SOEing is given in *Protocol 7.3*.

Primary PCR

The primary PCR amplifications are performed essentially according to the standard PCR protocol (*Protocol 2.1*) with the proviso that the highest possible annealing temperature should be used to prevent nonspecific amplification. The use of an efficient proofreading DNA polymerase, such as KOD DNA polymerase (Novagen), is recommended to reduce the possibility of errors and enhance the production of product and to prevent nontemplate directed addition of a 3'-nucleotide that might affect the subsequent steps. If you must use *Taq* DNA polymerase then the 5'-end of the primer(s) should be designed to be adjacent to a T in the sequence so that any addition of a 3'-A will not alter the DNA sequence of the final product. Two primary PCRs are performed simultaneously, one with primers P1 and P2 and the other with P3 and P4 to generate the two primary PCR products (*Figures 7.5* and *7.6*).

Secondary PCR

For the joining PCR amplification, the two amplified fragments from the primary PCRs should be mixed in a 1:1 molar ratio. The intention is to add the same number of molecules of each fragment. This is not the same as mixing the same volume or weight of each DNA fragment. For example, if fragment A is 600 bp long and fragment B is 300 bp then 100 ng of fragment A will contain the same number of DNA molecules as 50 ng of fragment B. If the gel used to estimate the concentration of the primary PCR products shows a good yield of a single product per reaction, then simply mixing aliquots of the reaction is appropriate when setting up the secondary PCR. Removal of the primers and other components is not really necessary, although a PCR clean-up step will not do any harm. However, if there is any doubt about the purity of the product it should be gel purified (Chapter 6).

In PCR 2 (*Figures 7.5* and *7.6*) primers P1 and P4 are also included. In the early cycles the two primary PCR products should anneal due to their over-lapping ends and self-prime to generate a double-stranded full-length molecule, which in subsequent cycles of PCR can be amplified exponentially by primers P1 and P4. This PCR should only be performed for about 15–20 cycles to reduce the possibility of errors being incorporated. Low yields from this PCR are best overcome by adding larger amounts of the two primary PCR products rather than by increasing the number of cycles used.

7.5 Point mutations

Point mutations refer to the replacement of one or more nucleotides within a sequence by other nucleotides (*Figure 7.7*). In the case of a protein coding sequence this usually has the effect of changing one or more amino acid codons so that the corresponding protein has one or more altered amino acids. In protein engineering the major advantage of site-directed muta-genesis is that a single amino acid can be changed anywhere in the protein, potentially altering the function or specificity of the protein, with minimal change in overall protein structure. For example a tyrosine codon (TAC) can be altered to an alanine codon (GCC) by changing T→G and A→C. Multiple mutations of a single codon can also be introduced by designing the primer to include a mixture of sequences at the codon position. For example the tyrosine codon (TAC) could be changed to alanine (GCC), glycine (GGC), serine (TCC) or cysteine (TGC) by using a mixed codon sequence of (G/T)(G/C)C; that is a mixture of G and T at position 1, G and C at position 2 and C at position 3. After mutagenesis, individual clones would be isolated and the mutation identified by DNA sequencing. More complex mixtures of mutations can also be introduced in a process usually called saturation mutagenesis. Theoretically, by using (A/G/C/T)(A/G/C/T)(G/C) at the target codon any of the 20 amino acids could be encoded, as well as one stop codon (TAG). In practice the situation is not quite so straightforward. There is some differential success in the coupling efficiencies of different nucleotides during chemical synthesis of oligonucleotides. Some codon sequences may there-fore be preferentially represented within the mixture thus creating some bias in the frequencies with which different codons may be isolated in a mutant plasmid population. Screening around 300 colonies from such a saturation library should allow all 20 replacements to be identified. Primers for intro-ducing point mutations into a sequence should usually be 20–40 nucleotides long depending upon the complexity of the mutations. The greater the number of mismatches being introduced, the longer the primer should be. For a single mismatch a primer of 20 nt should be sufficient, whereas for a saturation experiment at one codon a primer of 30 nt would be appropriate. For introducing 4 or 5 point mutations a primer of 35–40 nucleotides may be more appropriate. The primers for introducing mutations should correspond to primers P2 and P3 in the general SOEing scheme described above (*Figure 7.6*). In this case there should be around 10–15 nucleotides at the 5′-end of each primer that are perfectly matched to the template (*Figure 7.7*). These regions will give rise to the 3′-ends of the fragments used in the secondary PCRs so it is best to avoid mismatches in these regions.

7.6 Deletions and insertions

SOEing can also be used to introduce deletions (*Figure 7.8*) and insertions (*Figures 7.9, 7.10, 7.12*). The only prerequisite for all of these applications is the introduction, via a PCR primer, of an overlapping region of DNA at the ends of the two DNA fragments to allow them to be joined. Synthetic gene construction follows a similar strategy to join together segments of a gene that display overlapping terminal sequences (Section 7.13).

7.7 Deletion mutagenesis

The introduction of a deletion is a relatively simple task with SOEing as shown in *Figure 7.8*. Primers P2 and P3 are designed so that their 3′-regions correspond to sequences adjacent to the deletion end points. The primary PCRs with primers P1 + P2 and P3 + P4 will fail to copy the sequence destined for deletion. Primer P2 or P3 or both should also contain a 5′-extension sequence complementary to the other primer sequence. In the example shown (*Figure 7.8*), P3 contains a 5′ sequence complementary to P2. The ends of the two PCR products will therefore be complementary and can be joined in the PCR2 SOEing reaction to yield a product that now lacks the region targeted for deletion. This product can be readily cloned into a suitable vector. Deletions can also be introduced by inverse PCR as described in Section 7.2. The primers are simply designed to lie adjacent to the deletion end points.

7.8 Insertion mutagenesis

SOEing can be used to insert DNA sequences or for 'domain swapping' experiments. Different approaches are required depending upon the size of the inserted sequence and will be considered as (i) insertion of short sequences (1–80 nucleotides) and (ii) insertion of long sequences (>80 nucleotides).

Insertion of short sequences

If the insertion is to be added to the start or end of a target gene or coding region then a simple PCR amplification, with two primers P1 and P2 in which one primer carries the additional sequence, will achieve this outcome as previously discussed in Chapters 3 and 5. This strategy is frequently used for addition of restriction endonuclease cleavage sites or to add in-frame coding regions for detection and purification handles such as a 6 or 8 Histidine tag or a StrepTagII sequence or protease cleavage sites.

If the insertion is to be introduced within an existing sequence, or between two sequences that are to be joined, the most common approach is to use a SOEing strategy. The new DNA sequence can be added to the 5′-ends of the overlapping primers P2 and P3 (*Figures 7.9* and *7.10*).

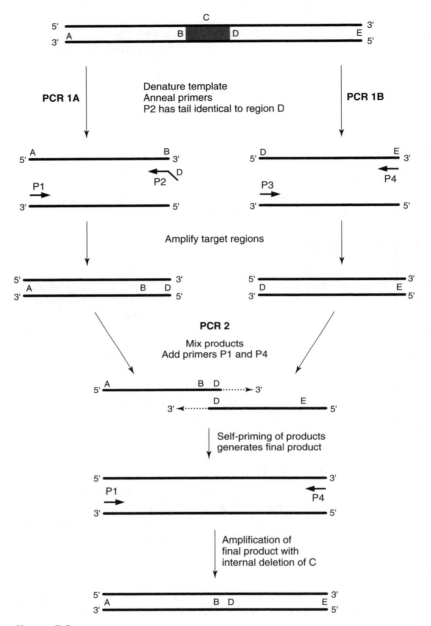

Figure 7.8

Schematic showing the deletion of a region of DNA by SOEing PCR. A gene with regions A–E is shown and it is intended to delete region C. In this case primer P2 is designed to anneal to region B and has a 5′-tail identical to region D. Two PCR reactions are performed on the template DNA, one using primers P1 and P2 and the other uses primers P3 and P4. The two PCR products therefore have region C absent, but have region D present on both products. This region allows annealing of the two products and self-priming in a further PCR. Amplification of the final product lacking region C is achieved by primers P1 and P4.

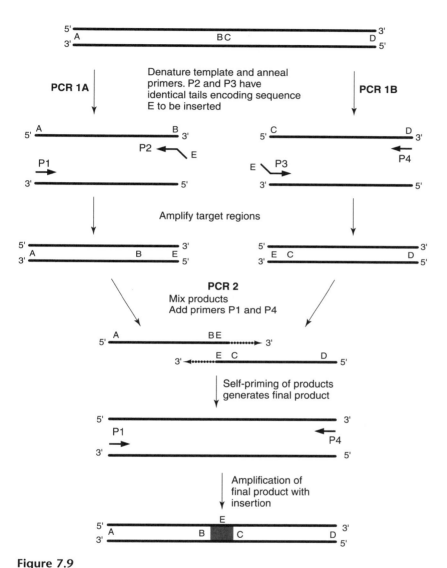

Figure 7.9

Schematic showing the insertion of a region of sequence E, between two existing sequences, B and C, in a gene. Primer P2 is designed to anneal to region B and to carry a 5'-tail representing all or part of region E. Primer P3 is designed to anneal to region C and to have a 5'-tail also representing all or part of region E, complementary to the sequence carried by primer P2. There must be at least a region of around 16 nucleotides complementary between the tails on P2 and P3 to allow the annealing of the PCR products and self-priming. Amplification of the full-length product is achieved by use of P1 and P4. The resulting product carries region E inserted between regions B and C. The size of the insertion that can be introduced is limited by the length of oligonucleotides that can be synthesized (see *Figure 7.10*); larger inserts can be introduced as described in *Figures 7.11* or *7.12*.

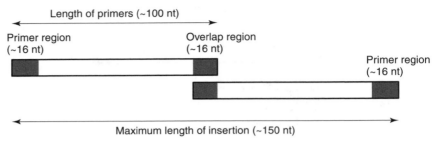

Figure 7.10

Length of insertion that can be introduced into a DNA sequence by SOEing using two overlapping oligonucleotides of 100 nucleotides in length.

The length of oligonucleotide that can be synthesized using current chemistry limits the insertion size. It is technically difficult to synthesize oligonucleotides longer than 100 nucleotides. Since the 3'-region (at least 16–20 nt) must act as an efficient PCR primer, the insertion sequence can only be around 80 nt. However, the SOEing strategy allows this size to be increased as the 5'-ends of primers P2 and P3 need only have sufficient overlap (around 15–20 nt) to allow them to anneal effectively and to prime DNA synthesis. So for two primers each 100 nt long, with 16 nt at the 3'-ends and 16 nt at the 5'-ends required for priming in the primary and secondary PCRs respectively, the maximum insert that could be introduced is $(2 \times 100) - (3 \times 16) = 200 - 48 = 152$ nt (*Figure 7.10*).

Insertion of long sequences

Before the advent of PCR, the insertion of long sequences into a target gene was generally dependent on conveniently situated restriction sites. PCR provides an elegant level of control over the generation of exactly tailored contructs. Here we describe two approaches for large insertions or replacements of DNA sequences by sticky-feet mutagenesis and SOEing.

Sticky-feet mutagenesis

'Sticky-feet' mutagenesis (4) can be used either as a method for inserting a new sequence or as a way of replacing a region of the existing sequence. Essentially, as illustrated in *Figure 7.11*, a region of DNA from gene X is amplified by using two primers that have 5'-extensions corresponding to the sequences of junction regions at which the insertion is to be made within the target gene Y. PCR with these two primers generates a DNA fragment with ends corresponding to the target gene. These 5'-ends act as 'sticky feet' that allow the PCR product to anneal precisely to the target gene. The 3'-end of the annealed fragment acts as a primer in a standard DNA synthesis reaction of the remainder of insert plus vector. The final stage, the plasmid replication step, does not involve a PCR amplification but rather is a linear amplification of the type previously discussed in the Quikchange process (Section 7.2). A method for sticky-feet mutagenesis is provided in *Protocol 7.4*, and uses a proofreading thermostable enzyme, such as KOD (Chapter 3) so that as soon as the long primer fragment anneals to the template it can be extended during the cooling period from the denaturation temperature.

The initial description of sticky feet mutagenesis used the single-stranded genome of M13 as template DNA for synthesis of a complete second strand of the circular genome. The template was prepared from a *dut⁻*, *ung⁻ E. coli* strain that leads to the incorporation of some dUTP instead of the normal dTTP in DNA synthesized within the cells. The *dut* gene encodes a dUTPase that maintains a low level of dUTP in the cells so the *dut⁻* mutation increases the cellular pool of dUTP. When dUTP is erroneously incorporated into DNA it is removed by the enzyme uracil *N*-glycosylase, encoded by the *ung* gene. This enzyme is not produced in the *ung⁻* background. So in the *dut⁻*, *ung⁻* strain dUTP levels are elevated, more dUTP is incorporated into DNA molecules, but these cannot be removed. This means that about 30 or so dUs are incorporated into each M13 template. Following primer-directed synthesis on the template, the newly synthesized strand contains no dU, but the original dU-containing template DNA strand is destroyed when the heteroduplex molecule is transformed into an *ung⁺* (uracil glycosylase-containing) *E. coli* host, leading to efficient selection of insertion mutants. While the use of filamentous phage such as M13 is becoming increasingly popular for the display of molecules on their surface in so-called phage display procedures, it is also true that they are not routinely used for many other cloning experiments. A benefit of this 'sticky feet' methodology is that it can be used to introduce insertions or sequence replacements without the need to clone PCR products. A method for this process is given in *Protocol 7.4*. By considering the approaches discussed in Section 7.2 it should be obvious that an alternative strategy to introduce mutations by sticky feet mutagenesis would be to use the *Dpn*I restriction selection used in the Quikchange procedures.

Primer design and primary PCR amplification

If a region of gene A is to be amplified and inserted into gene B then the primers used for 'sticky-feet' mutagenesis should contain:

- at the 3'-ends, a 15–20 nt region of gene A at the 3'-end; and
- at the 5'-ends a 20–30 nt region adjacent to the insertion site in gene B; these regions will fulfill the roles of 'sticky-feet'.

The length of the primers (*Figure 7.11*; primers 1 and 2) used to generate the 'sticky-feet primer' depends largely on the template source from which the primary PCR is being performed. In the case of complex templates, such as genomic DNA or cDNA libraries, long 3'-primer regions of 25–30 nucleotides are advisable to ensure specificity at high annealing temperatures.

Template preparation

If the UNG system is being used then a *dut⁻ung⁻* strain for propagation of the M13 or plasmid template leads to the incorporation of some uracils in place of thymidines (~30–40) in each template strand. This provides a

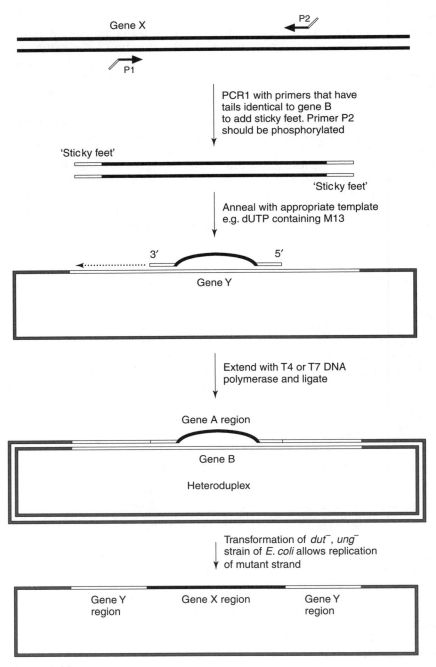

Figure 7.11

Sticky-feet mutagenesis. PCR is used to amplify a region of the target gene (gene X) with ends, defined by the primers, which are complementary to the gene Y into which the sequence is to be introduced. One primer, P2 in this case, must be phosphorylated so that the circular DNA synthesized by extension from the sticky-feet primer can be closed by ligation.

powerful selection method for mutant strands as the original template strand becomes a target for cellular repair mechanisms upon transfection into an *ung*⁺ strain with the mutant strand being used as the template for the repair process. This approach requires growth in a special strain and there can be some problems with template integrity during storage. A more robust approach would be to adopt the Quikchange approach which only requires that the template is grown in a common *dam*⁺ strain and the reaction subjected to a *Dpn*I digest before transformation.

Screening for mutants

Screening to identify mutants is best performed using PCR amplification and primers flanking the inserted DNA. This will identify clones carrying the inserted DNA as the fragment size will be larger than for clones that do not carry the insert. Alternatively use one flanking primer and a primer specific for the inserted sequence which will only yield a product from clones carrying the insertion.

Insertion of large fragments by SOEing

SOEing has been described for the introduction of short synthetic sequences of DNA earlier in this Section, but it can be adapted to introduce large segments of DNA. For example, in *Figure 7.12*, a segment of gene Y is to be inserted into gene X. Primers (P5 and P6) are designed to amplify the target segment from gene Y with 5′-tails that correspond to sequences for the target gene X, essentially as described for sticky-feet mutagenesis (see above). In parallel PCR reactions primers P1 and P2 and primer P3 and P4 are used to amplify the appropriate regions of gene X. It is important to have overlaps between the various fragments. So, P2 and P5 have complementary overlaps as do P6 and P3. After all three fragments have been purified by gel extraction or a PCR clean-up step there are two approaches for SOEing the fragments together. One possibility is that all three fragments can be mixed and two flanking primers (P1 and P4) can be included (*Figure 7.12*). As there is no intact wild-type gene present the flanking primers can only amplify the SOEn fragment.

Alternatively, one gene X fragment (P1 + P2 = fragment A–B) could be mixed with the insert fragment from gene Y (P5 + P6 = fragment –E–F–G–) and subjected to 1–5 cycles of PCR before addition of the other fragment from gene X (P3 + P4 = fragment C–D). The reaction can then be subjected to 1–5 cycles of PCR before the addition of the appropriate flanking primers (P1 and P4) for a further 10 or so cycles. Both approaches should yield the full-length insertion product.

7.9   Random mutagenesis

In site-directed mutagenesis the targeted sequence is usually identified due to sequence conservation or based on three-dimensional structural information to guide the choice of target residues likely to be important for structure or function. Such experiments have been of paramount importance in defining important aspects of regulatory DNA sequences and critical

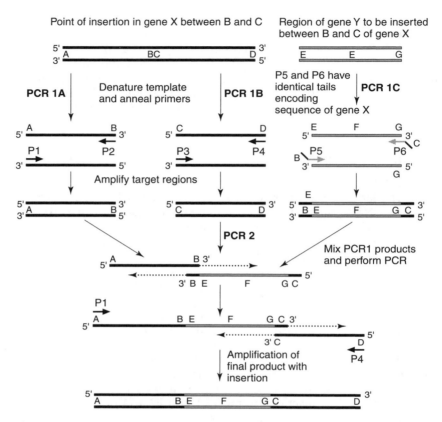

Figure 7.12

Insertion of a region (E–F–G) of gene Y within gene X between regions B and C. Three initial PCR reaction are performed. Primers P1 and P2 amplify region A–B of gene X; primers P3 and P4 amplify region C–D of gene X; primers P5 and P6 amplify region E–F–G of gene Y. These latter primers carry tails complementary to the insertion sites on gene X. Primer P5 carries a 5′-tail equivalent to region B and primer P6 carries a 5′-tail equivalent to region C. Combining the amplified products allows the construction of the hybrid gene. This could be performed by mixing all three PCR products, or, as shown in the scheme, by initially mixing products A–B and BE–F–GC and allowing self-priming; then adding product C–D and allowing self-priming; and then amplifying the full-length product with primers P1 and P4.

active site residues within receptor, enzymes and other proteins. Increasingly, however, it is clear that even with three-dimensional structural information it is not possible to identify sites that play critical roles in protein function, but which are distant from the active site. Random mutagenesis (5,6) and directed evolution are becoming increasingly important for (a) the generation of libraries of molecules carrying random mutations, and (b) the subsequent screening or selection of new variants that have distinct or improved properties compared with the wild-type starting molecule. PCR processes have an important role to play in the generation of libraries of random mutations. In Section 7.5 the concept of saturation mutagenesis was discussed. This represents a form of random mutagenesis since you are exploring the substitution of a single amino acid with any of the other 19 amino acids, although it is specifically targeting a particular amino acid. The approaches to random mutagenesis that will be highlighted in this section are more truly random in that they can lead to a change in any amino acid within a coding region.

7.10   PCR misincorporation procedures

You are already aware that *Taq* DNA polymerase has a relatively low fidelity in DNA replication, meaning that it will incorporate the wrong nucleotide once in every 10^4–10^5 nucleotides it synthesizes. In most PCR applications we are trying to reduce this error rate to ensure the template is copied accurately with no errors. However, for random mutagenesis by error-prone PCR we try to increase the rate at which the enzyme introduces errors. Cadwell and Joyce (5) identified conditions that led to such an enhanced error rate by altering the reaction conditions to favor misincorporation, leading to a significantly higher frequency of error rate during DNA synthesis. The key to this is the use of manganese and the alteration of the ratio of nucleotides concentrations. An error rate of approximately 0.05% or about one error in every 2000 nucleotides synthesized can be achieved.

The error rate may be further enhanced or controlled in longer products by including various nucleotide analogues that are rather promiscuous in their base pairing capabilities, thus favoring both transitions and transversions. Zaccolo and coworkers (7) described the use of dPTP (6-(2-deoxy-β-D-ribofuranosyl)-3,4-dihydro-8H-pyrimido-[4,5-C][1,2]oxazin-7-one) and 8-oxo-dGTP (8-oxo-2′deoxyguanosine) for error-prone PCR. dPTP can replace dTTP during DNA synthesis with only a four-fold reduction in incorporation efficiency, and leads to transversion and transition mutations A→G or T, T→C or G, G→A and C→T at between 1 and 4.4 per 100 bases synthesized. 8-oxodGTP cannot completely replace dGTP, but when a mixture of dGTP and 8-oxodGTP was used G→A, C→T, A→C and T→G mutations were observed at rates of 0.8–1 per 100 nucleotides synthesized. Clearly dPTP is the most effective, leading to mutation rates as high as 9.38%, while the combination of both compounds could generate mutations at an even higher frequency of 10.3%. There is now some difficulty in obtaining these nucleotide analogues commercially, but alternatives may be available or may be synthesized by a friendly chemist.

Random mutational libraries

The Quikchange approach can also be used to introduce random mutations based on error-prone PCR. For example Stratagene provide a system called GeneMorph with EZClone. This involves error-prone PCR of a region of the target gene defined by two opposing PCR primers by using an enzyme called Mutazyme II. The products of this reaction represent a population of molecules with a low level of sequence variation introduced during the error-prone PCR. These products are used as megaprimers in a Quikchange-type reaction. Both strands of the product act as primers on the plasmid template that carries the wild-type gene and in a linear amplification lead to new plasmid strands with single-strand ends corresponding to the megaprimers. These plasmid products will circularize by base pairing of the megaprimers, and following treatment with *Dpn*I to destroy the original plasmid template can be transformed into *E. coli* to generate a library of randomly mutated variants for the starting gene. The EZClone kit contains an enzyme mix to facilitate this plasmid replication. The process seems to work efficiently using a standard PCR reaction using a proofreading enzyme such as KOD DNA polymerase. The process is essentially as shown in *Figure 7.3* with the exception that a low level of mutations are introduced throughout the megaprimer sequence.

Spectrum of changes by error-prone PCR

Error-prone PCR provides an efficient mechanism for creating a library of variants that differ from the wild-type gene. Depending upon the level of mutations required, any single copy of the gene will contain one, a few or many nucleotide differences from any other copy of the gene. Some of the changes will not lead to any amino acid change in a protein coding sequence due to the degeneracy of the genetic code, while other changes will alter an amino acid. It is important to remember that this approach will allow the change of any one amino acid into only some, not all of the other 19. At these low rates of mutagenesis only one nucleotide in any codon will be changed, leading to a subset of possible amino acid substitutions. For example if single nucleotide changes were introduced into glutamic acid, GAA, it could give rise to:

AAA (lysine);
TAA (stop);
CAA (glutamine);
GGA (glycine);
GTA (valine);
GCA (alanine);
GAG (glutamate);
GAT (aspartate);
GAC (aspartate).

It would not give rise to any of the other 13 amino acids because in each case it would require more than one nucleotide to be changed per codon.

7.11 Recombination strategies

Variant libraries such as those generated by error-prone PCR can provide a useful source of new variant molecules, but such libraries may not allow the identification of combinations of mutations that display synergistic effects on enhancing protein function. It is possible to explore combinations of mutations that coexist within molecules by subjecting the DNA mutant library to some form of recombination, allowing beneficial mutations that may exist on separate molecules to be recombined so that they occur on the same DNA molecule. These processes can be termed 'recursive gene synthesis'. Recursive is defined as 'a data structure that is partially composed of other instances of the data structure'. As you will see from the following discussions this perfectly describes the processes that are involved in recombinational generation of sequence diversity.

Staggered extension process

Frances Arnold developed the staggered extension process (StEP) to achieve this controlled recombination (8,9). Having identified two or more useful variants from an error-prone PCR screen, these can be subjected to rounds of recombination, preferably under high-fidelity (proofreading) conditions to prevent accumulation of deleterious mutations (9). The effect is to generate combinations of the starting mutations that may prove beneficial. The library is usually small enough (up to 100 000 clones) to screen as individual clones, particularly if you have access to a robotic colony picker and liquid-handling robot. The process, illustrated in *Figure 7.13*, utilizes PCR with two flanking primers. The PCR conditions are quite standard, but the extension time is as short as possible. This usually means moving directly from the annealing to the denaturation temperature. In practice the polymerase will begin to extend from the primer even at the annealing temperature and so, as the reaction heats, DNA synthesis is occurring. The intention is to allow very short sections of DNA to be synthesized. At the next denaturation stage these short extension products will dissociate and at the next annealing step can anneal to a new template, perhaps containing a different mutation. A further short extension effectively recombines the short sequence copied from the original template sequence with a short sequence copied from the new template sequence. This process is called 'template switching' and this recursive synthesis is continued until full-length molecules are generated. Each new molecule should be the result of priming on several different templates and if these templates contained different mutations then some of these mutations should now exist on the same DNA molecules. The overall result is that some molecules may contain no mutations, some may contain one and others may contain several. The library can then be screened for the desired function of the protein.

The closer two mutations are in the sequence the less chance there is that they will be recombined. This molecular process shows segregation of mutations in a psuedo-Mendelian manner. Starting with two templates each carrying a mutation sufficiently far apart to allow recombination, the StEP process will yield new molecules in a 1:2:1 ratio carrying both, one and none of the mutations, respectively. The minimum amount of DNA that is synthesized during the short extension step, and therefore the minimum recombination distance, has been estimated to be around 34 nt. This was measured by analyzing the frequency with which mutations found on different DNA molecules could be recombined (9). The StEP approach has been demonstrated to generate variants with improved catalytic activity (8) and thermostability (9). The StEP process is generally not used as widely as DNA shuffling for generating recombinational libraries but, with the advent of new thermocyclers that have extremely rapid ramp rates for changing temperature, it is possible that there are some applications where

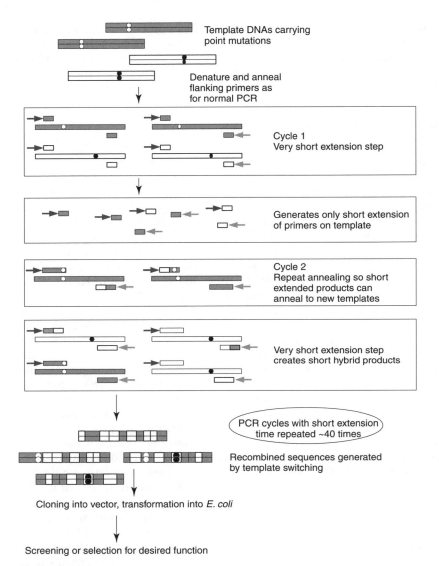

Template DNAs carrying
point mutations

Denature and anneal
flanking primers as
for normal PCR

Cycle 1
Very short extension step

Generates only short extension
of primers on template

Cycle 2
Repeat annealing so short
extended products can
anneal to new templates

Very short extension step
creates short hybrid products

PCR cycles with short extension
time repeated ~40 times

Recombined sequences generated
by template switching

Cloning into vector, transformation into *E. coli*

Screening or selection for desired function

Figure 7.13

Staggered extension process. DNA recombination is generated by the use of very short extension steps during rapid PCR cycles. Two versions of a target gene are colored gray and white to allow the process of recombination to be visualized. The two genes carry distinct mutations [○] and [●] selected as beneficial following error-prone PCR. Denaturation and annealing of primers to template strands using a very short extension time leads to short extended products. Repeated cycles of this rapid PCR lead to template switching and recombination at each PCR cycle. Eventually, after around 40 cycles of PCR, full-length recombinant genes are generated and amplified. Some carry no mutations, some have one of the mutations, others now carry both original mutations. After cloning into an appropriate expression vector, the library of variants is transformed into *E. coli* and clones are tested for the desired phenotype. Selected clones can be subjected to further rounds of error-prone PCR, StEP and screening.

this will prove useful and the recombination distance achievable may now be shorter than that measured originally.

DNA shuffling

Stemmer (10) described a PCR-based approach for the generation of large libraries of variants (*Figure 7.14*). In this case the method was originally designed for use with powerful selection such as antibiotic-resistant growth of bacteria. For example, if a variant confers the ability to grow on a higher level of an antibiotic then only the cells carrying this variant will grow whereas any that display normal or poorer resistance will die and will never be selected. Since this selection approach is more powerful than the screening approaches, it is possible to generate very large libraries where there will be many variants that are functionally worse than that encoded by the parent gene. The DNA shuffling procedure, illustrated schematically in *Figure 7.14*, uses physical fragmentation such as sonication, or controlled DNAse I digestion, to generate fragments from the starting DNA molecule. An example is shown in *Figure 7.15*. Generally these fragments will be in the range 50–100 bp and can be gel purified. The fragments are then subjected to about 40–60 cycles of a PCR but there are no flanking primers included. DNA synthesis therefore relies upon self-priming of annealed fragments of DNA. The random positions of ends of DNA molecules mean that various lengths of fragments will be synthesized and at the next cycle the extended products dissociate and can anneal to new fragments for a further round of self-priming extension. During this recursive PCR process the conditions favor incorporation of new mutations and these mutations can become recombined by DNA synthesis steps during the template switching.

Monitoring the reaction, by taking samples at stages during the reaction for gel analysis, reveals smears of DNA increasing in length with increasing cycle number indicating the formation of larger products with time (*Figure 7.15*). The final full-length products can be recovered by adding flanking primers and performing a limited number of standard PCR cycles followed by gel purification of the amplified band and cloning into an appropriate vector. A procedure for DNA shuffling is given in *Protocol 7.5*.

The shuffling approach is an extremely powerful tool for the generation and identification of new variants with improved properties such as catalytic rate, thermal and chemical stability, substrate specificity and binding capacity. These approaches represent a useful approach for the generation of variants of academic interest. However, for commercial applications patents are held by Maxygen Inc.

Family gene shuffling

In the original gene shuffling approach the starting material was a single gene and so error-prone PCR during the template switching steps that led to full-length gene products was an important mechanism for introducing diversity. An alternative strategy that relies upon naturally occurring variation between genes was then recognized to provide perhaps a more powerful approach for identifying improved proteins. In this approach a family of genes that encode proteins that have similar functions are used.

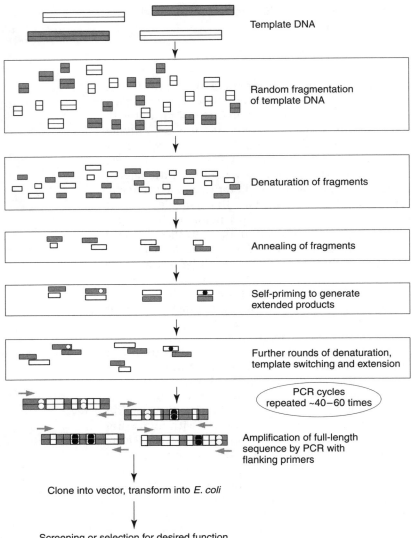

Template DNA

Random fragmentation
of template DNA

Denaturation of fragments

Annealing of fragments

Self-priming to generate
extended products

Further rounds of denaturation,
template switching and extension

PCR cycles
repeated ~40–60 times

Amplification of full-length
sequence by PCR with
flanking primers

Clone into vector, transform into *E. coli*

Screening or selection for desired function

Figure 7.14

DNA shuffling approach for generation of large variant libraries by random mutagenesis and
recombination. Two versions of a target gene are colored gray and white to allow the process of
recombination to be visualized. The DNA is fragmented by sonication or DNAse I treatment and a
subset of fragments of 50–100 bp are gel purified. Annealing of fragments by base pairing in a PCR
containing no primers results in self-priming and fragment extension. In addition the PCR conditions
lead to accumulation of new point mutations (○ and ●). Some 40–60 PCR cycles are performed to
reconstruct genes. Eventually full-length recombinant genes are generated; since this process is a
linear synthesis process, rather than an amplification reaction, it is often difficult to identify full-length
products. Addition of flanking primers allows the full-length products to be amplified by PCR. A
range of variants are generated due to new mutations arising during PCR and these are recombined
by the DNA shuffling process. After cloning into an appropriate expression vector, the library of
variants is transformed into *E. coli* and clones are tested for the desired phenotype. Selected clones
can be subjected to further rounds of DNA shuffling.

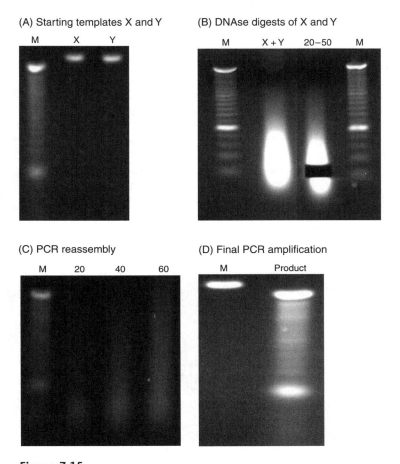

(A) Starting templates X and Y

(B) DNAse digests of X and Y

(C) PCR reassembly

(D) Final PCR amplification

Figure 7.15

Example of using DNA shuffling approach to recombine two genes. Gel (A): lanes X and Y; starting DNA fragments. Gel (B): X + Y, DNAse I digestion of mixed fragments from Gel (A) lanes X and Y; lane 25–50, a region of the gel containing DNA fragments in the region of 25–50 base pairs has been removed to recover the DNA fragments. Gel (C): PCR reassembly of gene from fragments isolated from Gel (B), lane 25–50. Lanes indicate products after 20, 40 and 60 cycles of reconstruction; an increase of smeared products can be observed with the increasing number of cycles. Gel (D): product is the PCR amplification of full-length gene product from Gel (C), lane 60, using flanking PCR primers. M; molecular size markers. Photographs kindly provided by D.J. Harrison.

These may for example be a family of a particular enzyme that catalyses the same reaction in a range of different organisms. Generally all of the starting molecules encode functional proteins, but sequence variation confers different properties on these proteins depending upon their evolutionary origins. For example some may be more efficient catalysts, others may be more thermostable or may use an altered spectrum of substrates.

In family shuffling the coding regions or those for specific domains are isolated, usually by PCR. These are then mixed and subjected to DNA shuffling by fragmentation of the DNA and regeneration of full-length sequence by the template switching recursive DNA synthesis procedure. In this case however, there are multiple different fragments derived from different genes and so template switching leads to recombination of sequences derived from different parental genes. However, since all the starting genes were functional, it is expected that many of the products will express hybrid proteins that retain some function. The question is whether the recombination process allows the acquisition of a range of beneficial properties from different parents, now within individual hybrid proteins.

Remarkable results have been achieved by DNA shuffling of families of genes. The first example was based on antibiotic resistance (11), which is quite easy to select for. However, more difficult screening strategies have been shown to be useful for isolating new functional proteins from family shuffling experiments. A particularly nice example demonstrates the power of this approach and the use of a heirarchical screening strategy for the selection of improved serine proteinases (12). There is also an informative study of the use of this type of family shuffling approach to begin to explore the functional importance of key residues within a protein, to begin tailoring substrate specificity, without any knowledge of protein structure. This involved the analysis of atrizine hydrolases (13).

7.12 RACHITT

Random chimeragenesis on transient templates (RACHITT) provides a new strategy for DNA recombination that leads to a higher level of crossovers than family shuffling (13). The method, illustrated in *Figure 7.16*, uses single-stranded molecules as the starting point. These can be generated by lambda exonuclease, a 5′-exonuclease that will degrade a strand of DNA containing a 5′-phosphate or 5′-overhang. The other DNA strand can therefore be protected by cutting with a 3′-overhang-generating enzyme. One strand contains deoxyuracil and acts as a scaffold for construction of the recombined complementary strands. Parental genes are used to prepare the complementary single strands to the scaffold and are then fragmented by DNAse I digestion. Following hybridization of these fragments and terminal primers to the scaffold any overhangs on the annealed DNA are trimmed back by the mixture of thermostable DNA polymerases included in the reaction. *Taq* DNA polymerase removes 5′-overhangs while *Pfu* DNA polymerase removes any 3′-overhangs and fills in any gaps. *Taq* DNA ligase then joins the fragments to create a full-length DNA strand. The scaffold strand is destroyed by treatment with uracil *N*-glycosylase and the recombined strands are then amplified by PCR. The parental genes could be a family of homologous genes, or a population of molecules generated by error-prone PCR. In addition mutations could also be introduced following the recombination step by error-prone PCR. The resulting amplified fragments are cloned into an appropriate expression vector for screening or selection of the desired variants.

7.13 Gene synthesis

PCR also provides an extremely efficient process for the generation of synthetic genes. It is increasingly common for genes to be resynthesized to

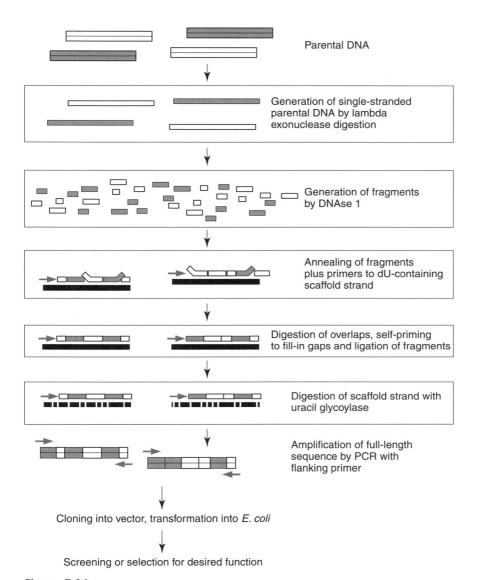

Parental DNA

Generation of single-stranded parental DNA by lambda exonuclease digestion

Generation of fragments by DNAse 1

Annealing of fragments plus primers to dU-containing scaffold strand

Digestion of overlaps, self-priming to fill-in gaps and ligation of fragments

Digestion of scaffold strand with uracil glycoylase

Amplification of full-length sequence by PCR with flanking primer

Cloning into vector, transformation into *E. coli*

Screening or selection for desired function

Figure 7.16

Schematic illustration of the RACHITT approach using recombination to introduce diversity. Complementary single-stranded scaffold DNA, containing dUTP, and parental DNAs are prepared by lambda exonclease digestion. The parental DNA is then fragmented by DNAse I digestion and annealed to the scaffold. Ends are trimmed and gaps repaired by a combination of *Taq* and a proofreading thermostable DNA polymerase, and nicks are sealed by thermostable DNA ligase. The scaffold strand is then digested by uracil N-glycosylase so that only the recombined strands remain. These can be PCR amplified by use of flanking primers and cloned into an appropriate vector, for transformation into *E. coli*.

allow the introduction of specific features such as convenient restriction sites and regulatory signals and to optimize the codon usage for expression of the encoded protein in a particular heterologous host. The strategy is illustrated in *Figure 7.17*. Oligonucleotides of 60–70 nucleotides can be produced and purified relatively easily and these can be designed to have overlaps with adjacent oligonucleotides. Pairs of oligonucleotides with

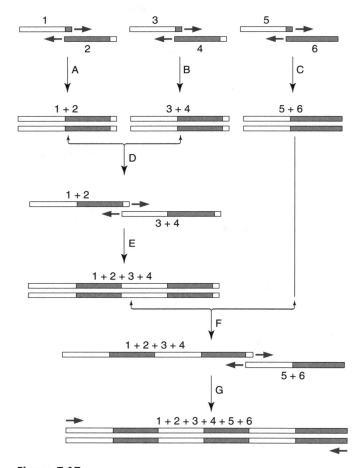

Figure 7.17

Example of total gene synthesis from 6 oligonucleotides of 60–70 nucleotides each. Oligonucleotide 2 overlaps with 3 by 12 nucleotides, and 4 overlaps with 5 by 12 nucleotides. Three separate self-priming fill-in reactions (A, B and C) are performed by one cycle of PCR denaturation, annealing and extension, to join respectively oligonucleotides 1 with 2, 3 with 4, and 5 with 6. An aliquot of the 1 + 2 product is then mixed with the product of 3 + 4 (D) and a further cycle of PCR is performed (E). This results in the joining of products 1 + 2 with 3 + 4 due to the 12-nucleotide overlap between oligonucleotidess 2 and 3. An aliquot of this product is then mixed with the product of 5 + 6 (F) and subjected to one cycle of PCR to allow a self-priming fill-in to generate the product 1+ 2 + 3 + 4 + 5 + 6 (G). The product can then be further amplified by PCR using either the original oligonucleotides 1 and 6 or smaller flanking primers.

overlaps of 10–15 nucleotides can be briefly heated and allowed to anneal and can then be extended by self-priming from their 3′-ends to generate a segment of the gene. Products of two such reactions that are adjacent in the gene can be mixed, denatured and allowed to anneal by virtue of the short overlapping sequences present on their ends. A further extension reaction will lead to a fragment now comprising four oligonucleotides. Depending upon the number of oligonucleotides that must be joined, this process can be continued with various combinations of oligonucleotides and extension products until the full-length gene has been generated and this can then be amplified by PCR using two flanking primers. The process is very rapid and simple. If the oligonucleotides have been designed carefully to avoid any possibility of mispriming on other oligonucleotides, there is no reason why all the oligonucleotides should not be mixed, subjected to a few rounds of heating, annealing and extension and then to around 5–10 cycles of standard PCR to generate a full-length gene.

It is important to ensure that the primers for use in gene synthesis are purified. The longer the oligonucleotides, the greater the proportion of erroneous sequences that will be present and the lower the yield of correct, full-length sequence that will be made. Purification of this product by HPLC or polyacrylamide gel electrophoresis is essential. Most manufacturers will provide such a purification service. Although this adds to the cost it significantly reduces the difficulties that are likely to be encountered otherwise in isolating the correct gene sequence. Even so, it is advisable to sequence several clones from a gene synthesis experiment to increase the likelihood that a correct clone will be identified.

Further reading

Arnold, F and Georgiou, G (eds) (2003) *Directed Evolution Library Creation*, Humana Press, Totowa NJ.
Arnold, F and Georgiou, G (eds) (2003) *Directed Enzyme Evolution: Screening and Selection Methods*, Humana Press, Totowa, NJ.

References

1. Hemsley A, Arnheim N, Toney MD, Cortopassi G, Galas DJ (1989) A simple method for site-directed mutagenesis using the polymerase chain reaction. *Nucleic Acids Res* **17**: 6545–6551.
2. Kim, Y-G, Maas, S (2000) Multiple site mutagenesis with high targeting efficiency in one cloning step. *Biotechniques* **28**: 196–198.
3. Horton RM, Hunt HD, Ho SN, Pullen JK, Pease LR (1989) Engineering hybrid genes without the use of restriction enzymes: gene splicing by overlap extension. *Gene* **77**: 61–68.
4. Clackson T, Winter G (1989) 'Sticky feet'-directed mutagenesis and its application to swapping antibody domains. *Nucleic Acids Res* **17**: 10163–10170.
5. Cadwell RC, Joyce GF (1992) Randomization of genes by PCR mutagenesis. *PCR Methods Appl* **2**: 28–33.
6. Hubner P, Iida S, Arber W (1988) Random mutagenesis using degenerate oligodeoxyribonucleotides. *Gene* **73**: 319–325.
7. Zaccolo M, Williams DM, Brown DM, Gherardi E (1996) An approach to random mutagenesis of DNA using mixtures of triphosphate derivatives of nucleoside analogues. *J Mol Biol* **255**: 589–603.

8. Moore JC, Jin H-M, Kuchner O, Arnold FH (1997) Strategies for the *in vitro* evolution of protein function: enzyme evolution by randon recombination of improved sequences. *J Mol Biol* **272**: 336–347.

9. Zhao H, Giver L, Shao Z, Affholter JA, Arnold FH (1998) Molecular evolution by staggered extension process (StEP) *in vitro* recombination. *Nature Biotechnol* **16**: 258–261.

10. Stemmer WP (1994) Rapid evolution of a protein *in vitro* by DNA shuffling. *Nature* **370**: 389–391.

11. Crameri A, Railland S-R, Bermudez E, Stemmer WPC (1998) DNA shuffling of a family of genes from diverse species accelerates directed evolution. *Nature* **391**: 288–291.

12. Ness JE, Welch M, Giver L, Bueno M, Cherry JR, Borchert TV, Stemmer WP, Minshull J (1999) DNA shuffling of subgenomic sequences of subtilisin. *Nature Biotechnol* **17**: 893–896.

13. Raillard S, Krebber A, Chen Y, Ness JE, Bermudez E, Trinidad R, Fullem R, Davis C, Welch M, Seffernick J, Wackett LP, Stemmer WP, Minshull J (2001) Novel enzyme activities and functional plasticity revealed by recombining highly homologous enzymes. *Chem Biol* **8**: 891–898.

14. Coco WM, Levinson WE, Michael J, Crist MJ, Hektor HJ, Darzins A, Pienkos PT, Squires CH, Monticello DJ (2001) DNA shuffling method for generating highly recombined genes and evolved enzymes. *Nature Biotechnol* **19**: 354–359.

Protocol 7.1 Inverse PCR mutagenesis

EQUIPMENT

0.5 or 0.2 ml PCR tubes

Thermal cycler

Gel electrophoresis tank

MATERIALS AND REAGENTS

Template plasmid DNA

Thermostable DNA polymerase and accompanying buffer

Phosphorylated oligonucleotide primers

2 mM dNTP solution

T4 DNA ligase and accompanying buffer

1 M Tris-HCl pH 8.5

Competent *E. coli* cells

1. Mix in a 0.5 ml microcentrifuge tube:
    *   2 µl 10 × polymerase buffer;
    *   2 µl 2 mM dNTPs;
    *   2 µl primer A (kinased; 5 pmol µl^{-1});
    *   2 µl primer B (kinased; 5 pmol µl^{-1});
    *   1–10 µl plasmid DNA (1–100 fmol);
    *   1 µl proofreading polymerase (2–3 units);
    *   Water to 20 µl.

2. Thermal cycle through:
    *   95°C for 2 min to denature the template; then 25 cycles of
    *   95°C for 1 min;
    *   55°C for 1 min;
    *   72°C for 5–6 min;
    *   and a final incubation at 72°C step for 10 min.

3. Purify the products from an agarose gel using an appropriate gel purification kit eluting in no more than 30 µl 10 mM Tris-HCl (pH 8.5) (Chapter 6).

4. Ligate 8–10 µl of the DNA solution in a 12–15 µl reaction containing T4 DNA ligase buffer including 1 mM ATP and 2

units T4 DNA ligase to catalyse recircularization of the linear plasmid molecules in a blunt-end ligation reaction.

5. Incubate at room temperature for 4–6 hours and transfect competent cells with 5–10 μl of the ligation reaction. Optionally a *Dpn*I digest could be added at this point to ensure that any wild-type template molecules which will be methylated are destroyed.

NOTE

The annealing temperature will depend upon the melting temperature of the oligonucleotides and it should ideally be as high as possible to ensure specificity.

Protocol 7.2 Quikchange mutagenesis of plasmid DNA

EQUIPMENT

0.5 or 0.2 ml PCR tubes

Thermal cycler

Gel electrophoresis tank

MATERIALS AND REAGENTS

Template plasmid DNA

KOD DNA polymerase and accompanying buffer (Novagen)

Complementary oligonucleotide primers (~35 nt long with central mutation(s)) 125 ng μl⁻¹

2 mM dNTP solution

25 mM MgSO₄

1 M Tris-HCl pH 8.5

Competent *E. coli* cells

1. Mix in a 0.5 ml microcentrifuge tube:
   - 5 μl 10 × polymerase buffer;
   - 5 μl 2 mM dNTPs;
   - 2 μl 25 mM MgSO₄;
   - 1 μl primer A (125 ng);
   - 1 μl primer B (125 ng);
   - 1–10 μl plasmid DNA (20–50 ng);
   - 1 μl KOD DNA polymerase (2–3 units);
   - Water to 50 μl

2. Thermal cycle through:
   - 95°C for 2 min to denature the template; then 25 cycles of
   - 95°C for 30 s;
   - 55°C for 30 s;
   - 72°C for 1–2 min kb⁻¹ of template (for example for a 5 kb plasmid up to a 10 min extension step should be used)[1];
   - and a final incubation at 72°C step for 10 min.

3. Remove an aliquot (~5 μl) as a control for gel analysis.

4. Add 1 μl *Dpn*I (New England Biolabs) and digest at 37°C for 1–2 h.

5. Remove a 5 µl aliquot and run together with the DNA from step 8, on a 1% agarose gel. As an additional control an aliquot of the original plasmid template can be subjected to *Dpn*I digestion and separated on the same gel. If the reaction has worked there should be a substantial band in both the *Dpn*I undigested and digested samples, indicating the production of a product that is resistant to *Dpn*I digestion[1]. If you use a plasmid template control this should be digested by *Dpn*I[2].

6. Transform a 4 µl aliquot of the digested reaction into suitable competent *E. coli* cells.

7. Screen around 4 colonies to identify mutants by DNA sequencing. Generally 2–4 clones will be mutant.

NOTE

1. If the results show a smeared and high molecular weight product this can be due to too high a concentration of template DNA or sometimes to the extension step being too long. Try reducing the amount of template and then reducing the time of the extension step to 1 min/kb.

2. The gel analysis step is optional. You could simply proceed to the transformation step after the *Dpn*I digestion; however, the additional time to analyse the products on a gel can provide confidence that the reaction has worked efficiently.

Protocol 7.3 Splicing by Overlap Extension (SOEing)

EQUIPMENT

0.5 or 0.2 ml PCR tubes

Microcentrifuge

Thermal cycler

Gel electrophoresis tank

MATERIALS AND REAGENTS

Plasmid template

PCR reagents

Oligonucleotide primers

Ethanol

1 M Tris-HCl pH 8.0

0.8% agarose (100 ml; 0.8 g agarose in 100 ml of 1 × TAE)

Part A – Primary PCRs

1. Set up two PCRs to perform primary PCR amplifications of the two DNA fragments to be spliced. One reaction should contain primers 1 and 2 and the second primers 3 and 4. PCR conditions and the reaction mixes should be as described in *Protocol 2.1*, except use:
   - 100–500 ng of plasmid template and only 10–15 cycles;
   - an annealing temperature as close to 72°C as possible.

2. Analyze the two amplified products on an agarose gel. Depending on the purity of the products either:
   - purify using a PCR clean-up kit (Chapter 6) to remove buffer and primers;
   or
   - concentrate the samples by ethanol precipitation, separate through a 0.8% agarose gel, excise the two fragments and recover from the agarose (Chapter 6).

3. Elute each fragment in a low volume (20–30 μl) of 10 mM Tris-HCl (pH 8.0) and estimate the concentration on an

agarose gel using known quantities of appropriate DNA molecular size markers.

Part B – Secondary PCR

1. Mix the fragments in a 1:1 molar ratio so that the amount of DNA added to PCR 2 is around 100 ng. Add 25 pmol each of primers 1 and 4.

2. Perform PCR 2 under the same temperature regime as for PCR 1 for 10–15 cycles.

3. Size fractionate the final product through an agarose gel, gel purify the fragment and clone into a suitable vector (Chapter 6).

Protocol 7.4 'Sticky-feet' mutagenesis

EQUIPMENT

0.5 or 0.2 ml PCR tubes

Thermal cycler

Gel electrophoresis tank

MATERIALS AND REAGENTS

Phosphorylated oligonucleotide primers

PCR reagents

Phagemid single-stranded template DNA

5 mM dNTP solution

Thermostable DNA polymerase and accompanying buffer

T7 DNA polymerase and accompanying buffer

100 mM DTT

T4 DNA ligase and accompanying buffer

500 mM EDTA pH 8.0

0.8% agarose (100 ml; 0.8 g agarose in 100 ml of 1 × TAE)

1. Make sure that the appropriate primer for the primary PCR amplification contains a 5'-phosphate prior to use. This can be requested when the primers are synthesized or you can perform a T4 DNA kinase reaction (*Protocol 3.1*).

2. Perform a PCR amplification with the two primers, as described in *Protocol 2.1*. Use an annealing temperature as close to the polymerization temperature as possible.

3. Size-fractionate the reaction products through an agarose gel and purify the 'sticky-feet' fragment (Chapter 6).

4. Set up a DNA extension reaction containing:
   - ~ 100–200 pmol 'sticky-feet primer';
   - 200 ng of M13 or phagemid single-stranded template DNA propagated in a *dut⁻, ung⁻* strain, or alkali denatured and neutralized supercoiled plasmid;
   - 5 mM each dNTP;
   - 10 × DNA polymerase buffer;

- 2–5 units of a thermostable proofreading DNA polymerase (Chapter 3).

5. Incubate the reaction at
   - 94°C for 2 min;
   - 65°C for 1 min;
   - 37°C for 1 min;

   with a ramping rate that allows about 1 min between the temperatures, to allow template denaturation and annealing of the sticky feet fragment.

6. To the reaction add the following components in order:
   - T7 DNA polymerase buffer;
   - 5 mM ATP;
   - 100 mM DTT;
   - 400 units T4 DNA ligase; and
   - 1 unit T7 DNA polymerase.

7. Incubate the reaction at 37°C for 30 min, inactivate with 1 μl of 500 mM EDTA and transform a *dut⁺, ung⁺ E. coli* strain with 1/10 of the final reaction.

8. Screen for mutant plaques (M13) or colonies (phagemid/plasmid) using PCR amplification and primers flanking the inserted sequence. While M13 phage are plated for the production of plaques representing the slower growth of cells infected by the phage, phagemids can infect a cell but then replicate as plasmids and carry none of the phage infection functions. Phagemid transformants are therefore selected for as colonies capable of growth on plates containing an antibiotic for which the phagemid carries a resistance gene.

Protocol 7.5 DNA shuffling

EQUIPMENT

1.5 ml microcentrifuge tubes

0.5 or 0.2 ml PCR tubes

Adjustable heating block or water bath

Thermal cycler

MATERIALS AND REAGENTS

Template DNA

$10 \times$ DNAse I digestion buffer (400 mM Tris-HCl (pH 7.9), 100 mM NaCl, 60 mM MgCl$_2$, 100 mM CaCl$_2$)

DNAse I

Ethanol

4% Nusieve GTG agarose (100 ml; 4 g Nusieve GTG agarose in 100 ml 1 $\times$ TAE)

Thermostable DNA polymerase and accompanying buffer

Flanking oligonucleotide primers

2 mM dNTP solution

This procedure describes shuffling of a single gene or several gene fragments.

PART A – TEMPLATE DNA

1. Prepare template fragments either by PCR amplification or by restriction digestion. It is necessary to use approximately 0.5 μg of template DNA in a shuffling reaction. Since only a proportion of the starting DNA will be fragmented in the correct size range you must ensure you generate about 5 times more of each template (approximately 2.5 μg per template fragment)

2. Purify the fragments from vector and if PCR is used, it is essential to remove the primers. Use an appropriate purification approach such as a spin column or gel purification (Chapter 6).

PART B – DNASE I DIGESTION

3. Mix the following components in a microcentrifuge tube:
   - 5 µl 10 × DNAse I digestion buffer;
   - 5 µg template DNA;
   - 0.5 units DNAse I;
   - total volume 50 µl.

4. Incubate at room temperature for 15 min.

5. Terminate the reaction by heating to 90°C for 10 min.

6. Ethanol precipitate (Chapter 6).

PART C – FRAGMENT PURIFICATION

7. Electrophorese the DNA through a 4% NuSeive GTG agarose gel (BioMolecular Whittaker) using TAE buffer with EDTA at 0.2 mM.

8. Visualize the ethidium bromide stained gel under UV illumination and recover the appropriate size fraction (normally 50–100 bp, although for small genes recovering a smaller size range such as 25–50 bp is recommended).

9. Trim the slice to remove excess agarose, but minimize exposure to UV.

PART D – DNA SHUFFLING REACTION

10. Combine in a 0.5 ml microcentrifuge tube:
    - gel slice maximum 12.5 µl;
    - 5 µl 10 × *Taq* DNA polymerase buffer;
    - 5 µl 2 mM each dNTP;
    - 1–3 units *Taq* DNA polymerase;
    - total volume 50 µl.

11. Perform thermal cycling (for an instrument-monitoring tube temperature) according to the following scheme:
    - 95°C for 5 min; then 40–60 cycles of
    - 95°C, 30 s;
    - 50°C, 30 s;
    - 72°C, 1 min;
    - and a final 72°C, 5 min extension.

12. Purify the final product using an appropriate spin column method as the reaction contains agarose eluted in approximately 30 µl 10 mM Tris-HCl (pH 8.0).

PART E – FINAL PCR STEP

13. Combine in a microcentrifuge tube:
    - 1.25 µl shuffled DNA;
    - 5 µl 10 × *Taq* DNA polymerase buffer;
    - 5 µl 0.2 mM each dNTP;
    - 25 pmol each flanking primer;
    - 1 unit *Taq* DNA polymerase;
    - total volume 50 µl.

14. Perform thermal cycling under the same conditions in step 11 for 20–30 cycles. If insufficient full-length product is generated then the number of cycles can be increased.

15. Clone the products into an appropriate expression vector for subsequent screening or selection.

Protocol 7.6 Gene synthesis

This Protocol describes the synthesis of a gene of approximately 300 base pairs from 6 oligonucleotides each 65–70 nt long and with 12–15 nt terminal overlaps.

EQUIPMENT

Microcentrifuge

Microcentrifuge tubes

Thermal cycler

Agarose gel system

MATERIALS AND REAGENTS

Oligonucleotides designed to generate synthetic gene at 50 ng μl^{-1}

Flanking primers to amplify the full-length product

Proofreading thermostable KOD DNA polymerase

10 × KOD polymerase buffer

2 mM dNTPs

25 mM $MgCl_2$

Water

1. Set up three parallel reactions, each containing one primer pair with reference to *Figure 7.17*: (primer 1 + 2); (primer 3 + 4); (primer 5 + 6):
   - 3 µl primer A (1 or 3 or 5);
   - 3 µl primer B (2 or 4 or 6);
   - 1 µl 10 × buffer;
   - 1 µl 2 mM dNTPs;
   - 0.4 3 µl 25 mM $MgCl_2$;
   - 1 unit polymerase;
   - total volume 10 µl.

2. Mix the reactions by a brief (1 s) centrifugation step

3. Place in a thermocycler and incubate as follows:
   - 94°C, 2 min;
   - 25°C, 2 min;
   - 72°C, 2 min.

   This step allows the annealing of the oligonucleotides and their self-extension to generate complete double-stranded fragments.

4. From each reaction take a 5 μl aliquot and combine in a fresh tube[1].

5. Repeat the incubation cycle in step 3.

6. Set up the following reaction:
   - 15 μl reaction mix from step 5;
   - 1 μl flanking primer 1 (100 pmol 15 μl^{-1});
   - 1 μl flanking primer 2 (100 pmol μl^{-1});
   - 3.5 μl 10 × polymerase buffer;
   - 3.5 μl 2 mM dNTPs   ;
   - 1.4 μl 25 mM MgCl$_2$;
   - 1 μl KOD DNA polymerase;
   - total volume 50 μl.

7. Perform thermal cycling according to the following scheme:
   - 94°C, 15 s;
   - 60°C, 15 s;
   - 72°C, 15 s to 1 min depending on size of gene.

8. Analyze a 1–5 μl sample on a 2% agarose gel to determine whether a correct sized product is visible.

9. Set up a restriction digest if required to digest the product to allow cloning into the appropriate cloning vector.

10. Separate the sample on a 2% agarose gel, excise the correct band and purify using an appropriate gel purification procedure (Chapter 6).

NOTES

1. At step 4 it is possible to perform additional steps to join pairs of fragments together in a stepwise manner. For example, you could set up a reaction to join product (1 + 2) to product (3 + 4) according to step 3. Then an aliquot of this reaction (1 + 2 + 3 + 4) could be added to an aliquot of the product (5 + 6) reaction and joined according to step 3. Finally in step 6 the full-length gene would be amplified.

Analysis of gene expression

8

8.1 Introduction

The varied phenotypes observed for both unicellular and multicellular organisms result from differences in the genes and alleles that comprise the genomes of each species. However, most cell types of a multicellular organism, such as nerve cells, liver cells, bone cells and blood cells, also show striking phenotypic variations. Similarly plant development is governed by differential expression of genes in different tissues and cell types. The DNA sequence of the genome in all cells is identical, but changes in the methylation state of regions of the genome and regulation of transcriptional processes leads to differential expression of cell-specific genes during development. In modern biology, accurate analysis of gene expression has become increasingly important not only in improving our understanding of gene and protein functions but also to detect low-level transcripts as part of biotechnological applications or in medical diagnosis (1). The website http://www.cs.wustl.edu/~jbuhler/research/array/#cells contains a useful introduction to comparative gene expression analysis.

For many years the conventional approaches to analyzing gene expression have been by Northern blot, *in situ* hybridization or RNAse protection assays. While these are still used extensively, they are often time consuming and are relatively insensitive, making detection of rare transcripts difficult or impossible. The development of PCR as a tool for analysis of gene expression patterns and to detect rare transcripts has revolutionized the sensitivity of gene expression analysis. It is now possible, using fluorescent dyes, to perform real-time analysis of accumulation of multiple products to provide more sophisticated information on relative levels of different gene transcripts. Changes in gene expression of even more genes can be analyzed in parallel by the use of microarrays which can allow several tens of thousands of gene probes to be investigated in a single experiment. This Chapter outlines how PCR can be used to analyze gene expression patterns and will describe current PCR techniques that allow quantitative gene expression analysis, and cellular and subcellular detection of transcript levels. A major technology for analysis of differential gene expression is real-time PCR and Chapter 9 has now been devoted to this important topic.

8.2 Reverse transcriptase PCR (RT-PCR)

Analysis of gene expression requires accurate determination of mRNA levels. But PCR is based on amplification of DNA rather than RNA, so how can it be used for mRNA analysis? The answer is that first, mRNA is converted into DNA using the well-known process of reverse transcription, which is used by RNA viruses to convert their genomic RNA into a DNA within the host cell; and second, PCR amplification is performed on the resulting complementary DNA (cDNA).

Standard RT-PCR

Standard RT-PCR offers a rapid, versatile and extremely sensitive way of analyzing whether a target gene is being expressed and can provide some semi-quantitative information about expression levels. Theoretically RT-PCR should be able to amplify one single mRNA molecule, although in practice this is not likely to be a realistic goal. However, RT-PCR is an extremely valuable tool when limited material, such as specific differentiated cells, is available. In this context RT-PCR can be used either to detect specific transcripts by using sequence-specific primers, or to create cDNA libraries by using generic primers such as oligo-dT and either random oligonucleotides or 5'-cap-specific primers such as the SMART II oligonucleotide (Clontech). The following Sections describe the steps involved in RT-PCR.

The reverse transcriptase reaction

RT-PCR is based on the ability of the enzyme reverse transcriptase, an RNA-dependent DNA polymerase, to generate a complementary strand of DNA (first-strand cDNA) using the mRNA as a template. The reverse transcriptase reaction can be performed on either total cytoplasmic RNA or purified mRNA. It is important that no genomic DNA is present, as this will also provide a template for the PCR amplification step. An appropriate control for any contaminating DNA is a control reaction in which the reverse transcriptase step is omitted. Many commercial kits generate high-quality DNA-free total or mRNA preparations, or an RNAse-free DNAse I digestion step can be included in the RNA extraction protocol. To analyze a previously characterized gene the primers can be designed to amplify across an intron, thus allowing simple identification of contaminating genomic DNA that will contain the intron while the transcript will not. This means DNA will give rise to a longer product than the RNA transcript. The method is sometimes called intron-differential RT-PCR. The use of purified mRNA is recommended since this generally gives rise to a higher yield of first-strand cDNA. When analyzing low abundance transcripts the use of purified mRNA is important for success since the relative concentration of the target mRNA will be much higher than when using total cytoplasmic RNA. A wide variety of simple-to-use kits, based on the use of oligo-dT annealing to the 3'-polyA tract of eukaryotic mRNAs, are available for purifying mRNA (*Figure 8.1*).

The next step is to copy the mRNA to first-strand cDNA (*Figure 8.2*). This is often done using an oligo-dT primer that can anneal to the 3'-polyA tail of eukaryotic mRNAs and allows reverse transcriptase to synthesize cDNA from each mRNA molecule present in the reaction. This can be carried out either using purified eluted mRNA or purified mRNA still attached to a solid support matrix. There are two common types of reverse transcriptase; Avian

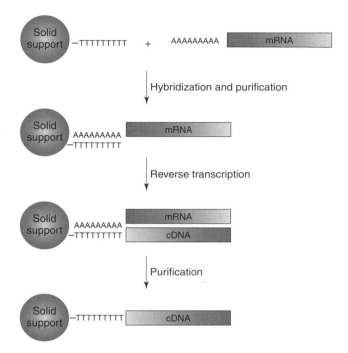

Figure 8.1

mRNA purification using an oligo-dT solid support matrix and subsequent first-strand cDNA synthesis.

Myeloblastoma Virus (AMV), reverse transcriptase and Moloney Murine Leukemia virus (M-MLV) reverse transcriptase. Versions that allow efficient copying of long mRNAs are available, for example the M-MLV RNase (H–) that carries a point mutation (Stratagene, Promega). However, new enzymes are being produced such as the *Carboxydothermus hydrogenoformans* polymerase (Roche Applied Science), which displays reverse transcriptase activity at a high reaction temperature between 60°C and 70°C. AMV-RT has both 5′→3′ primer-dependent polymerase activity with either RNA or DNA as template and a 3′→5′ RNAse H activity that degrades the RNA portion of the RNA-DNA heteroduplex product of cDNA synthesis. The M-MLV-RT is essentially identical to the AMV enzyme but it can only use RNA as a template.

For a standard first-strand cDNA reaction using AMV-RT approximately 1 μg of total RNA or 10–100 ng mRNA should be used. Depending on the abundance of the target mRNA species, the optimal amount of RNA may need to be determined empirically by testing various starting amounts. A standard reverse transcriptase reaction is described in *Protocol 8.1*.

Usually first-strand cDNA synthesis is very reliable and an aliquot of the reaction can be taken immediately for PCR amplification. However, if there is any doubt about the quality of the mRNA or the cDNA synthesis reaction, or you fail to obtain a PCR product, the success and efficiency of the reverse transcriptase reaction should be monitored. For example, if no PCR amplified product is detected it is important to know whether this is due to the failure of the first-strand cDNA synthesis reaction or of the PCR reaction. If a gene-specific primer to be used for the subsequent PCR step lies close to the 5′-end of the gene, it is useful to know that reverse transcription has yielded first-strand cDNA of appropriate length. If the abundance of the transcript is extremely low it may be necessary to optimize the first-strand cDNA synthesis conditions by varying mRNA and primer concentrations/combinations. The simplest way of analyzing the efficiency of first-strand synthesis is to substitute one of the nucleotides with a radiolabeled nucleotide, such as [α–^{32}P] dATP or dCTP that will be incorporated into the cDNA, and then to calculate the final incorporation value by scintillation counting. A less quantitative method, but one that provides information on the size range of cDNA products, is gel electrophoresis. An aliquot of the first-strand cDNA reaction can be fractionated through an agarose gel after RNAse digestion to remove the template RNA. The first-strand synthesis product will consist of single-stranded DNA so cannot be visualized efficiently using ethidium bromide. However, the radiolabeled cDNA can be analyzed by autoradiography of the gel. This can be done directly by covering the gel in plastic film and then exposing it to X-ray film or a phosphorimager plate. Alternatively, the gel can be transferred to a membrane, such as nitrocellulose, by using standard Southern blot procedures (Chapter 5) and the membrane can be exposed to film or an imager plate. If radiolabel was not included, the membrane or the agarose gel can be stained using a single-stranded specific nucleic acid dye such as SYBR® Green II nucleic acid gel stain (Molecular Probes) or Fast RNA Stain′ (HealthGene Corporation). A successful first-strand cDNA synthesis reaction produced by oligo-dT priming should appear as a smear from a position greater than 2 kb due to the heterogeneous mixture of cDNA products. RNA markers can be used to help assess the size range of cDNA products. Once the success of the first-strand cDNA reaction has been verified the remainder of the reaction products can either be used directly for PCR or stored at –80°C.

The analysis of RT-PCR amplification products is performed by the detection methods described in Chapter 5. Despite the possibility of low levels of amplification due to low initial concentrations of target transcript, standard agarose gel electrophoresis and ethidium bromide staining is usually sufficient to detect the final RT-PCR amplification product.

The PCR reaction

The next step is to amplify the cDNA by PCR as described in *Protocol 2.1*. Appropriate upstream and downstream primers are used and can either be specific to the target gene, or, for cDNA library construction, generic. Due to the single-stranded nature of the first strand cDNA, the early cycles of the PCR involve linear amplification as the first strand can only act as template for one of the primers. Exponential amplification from both primers occurs once sufficient copies of the second strand have been generated. In practice this has no effect on the final PCR amplification yield.

In some cases, particularly when transcript levels are low, some optimization of PCR conditions will probably be necessary to obtain a

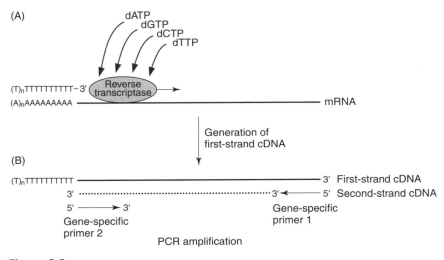

Figure 8.2

Diagram showing (A) reverse transcription from mRNA using an oligo-dT primer and (B) second-strand cDNA synthesis.

convincing result. The optimization can of course be performed on the first-strand cDNA material but, if extensive optimization is required, this will be very wasteful and will require the use of large amounts of reverse transcriptase.

A more economical way of optimizing the PCR parameters is to use the same reaction components as for the RT-PCR itself but using genomic DNA or plasmid DNA containing either the genomic region or the cDNA of interest. Of course by using double-stranded DNA for the optimization experiments, the PCR conditions are not strictly mimicked, but should allow you to determine the best temperature profiles and primer combinations for any given sample.

8.3   Semi-quantitative and quantitative RT-PCR

While standard RT-PCR can detect the presence or absence of mRNA species it does not provide a quantitative measurement of levels of gene expression principally due to the 'plateau effect' described in Chapter 2. However, by modifying the standard method RT-PCR can be used to quantify the levels of mRNA in a sample or provide insight into the relative expression levels between different cell types or in response to external stimuli.

Semi-quantitative RT-PCR

If relative differences in transcript levels are to be compared between different cell types, a semi-quantitative approach may be sufficient. The simplest way of performing such analysis is to determine the amounts of PCR product during the exponential phase of the PCR but before the plateau phase (Chapter 2). While this approach does not give any absolute value of the mRNA level in your starting sample it will readily detect differences of 10–20-fold in mRNA levels between different samples. This method can be useful for analyzing changes in the level of a target transcript in identical tissue or cells in response to external stimuli. Of course valid comparisons are only possible when the same primer combinations and reaction conditions are used for all samples. The PCR experiments should be performed in parallel at least twice to ensure that the results obtained are consistent and reproducible. An example of a semi-quantitative analysis analyzed by agarose gel electrophoresis is shown in *Figure 8.3*.

An oligo-dT primer should be used for the first-strand cDNA synthesis because eukaryotic mRNA molecules have a polyA tail, ensuring that the level of cDNA synthesis reflects the level of the starting target mRNA. The recommended way of determining the efficiency of cDNA synthesis is to measure the incorporated level of radiolabeled nucleotides by scintillation counting. Identical quantities (radioactivity counts per minute) of each first-strand cDNA reaction should be used for PCR (Section 2.1). Aliquots should be removed from each reaction during the PCR every 3–5 cycles for the first 15–20 cycles. This ensures that the reaction is being sampled during the exponential phase of the PCR and that the plateau is never reached. Agarose gel electrophoresis may not be sufficiently sensitive to detect slight differences in amplification levels between samples. In such cases Southern blot analysis (Chapter 5) should be performed using either a DNA or an oligonucleotide probe. For the detection of slight differences between high abundance mRNA species it may be necessary to perform serial dilutions of the RNA or PCR products to achieve the optimal range for accurate estimation of mRNA levels. For this purpose dot-blot analysis is recommended as large numbers of samples can be analyzed simultaneously. The measure-

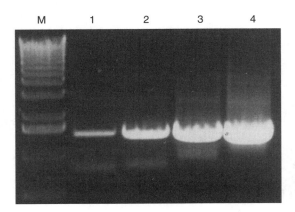

Figure 8.3

Agarose gel showing semi-quantitative RT-PCR analysis from plant RNA. Amplification was performed using primers designed for an abundantly expressed root gene. Lanes 1 and 2 represent RT-PCR of RNA from *Arabidopsis thaliana* flowers using 10 (lane 1) and 15 (lane 2) amplification cycles. Lanes 3 and 4 represents RT-PCR of RNA from *Arabidopsis thaliana* roots using 10 (lane 3) and 15 (lane 4) amplification cycles.

ment of signal intensities can either be performed by densitometry measurements of X-ray films or by a phosphoimager. X-ray film has a major limitation since even short exposures to different amounts of PCR product can appear equally intense due to the nonlinear nature of X-ray film. However, it can be useful if the amounts of PCR product are strikingly different. If possible use a phosphoimager, as even small differences in signal intensity can be accurately determined. If you do not have access to a phosphoimager an alternative is scintillation counting of isolated products on sections of the filter.

Virtual Northern blotting

Semi-quantitative analysis of gene expression profiles, either by Northern blot analysis or by differential display (Section 5), can lead to apparent false expression patterns and so it is best to perform an experiment based on an alternative approach to verify the result. For example a differential display result could be confirmed by Northern blot analysis or a Northern blot result could be confirmed by an RNAse protection assay. However, the bottleneck for such approaches is the requirement for microgram amounts of RNA. To overcome this problem of the availability of material a new approach involving an intrinsic PCR 'amplification' step has been incorporated into the Northern blot procedure creating a virtual Northern blot. The approach was first described by Clontech and has now been used successfully in place of standard Northern blot analysis. The principle is to generate full-length double-stranded cDNA and to incorporate an amplification step to boost the measurable levels of 'transcript' in the form of cDNA. This process requires between 50 and 500 ng of total RNA, which is significantly less than is required for standard Northern blotting (2–10 µg). Clontech's SMART™ PCR cDNA synthesis kit facilitates production of high-quality cDNA from total or polyA RNA as described more fully in Chapter 10 (Section 10.1).

In order to allow for semi-quantitative analysis it is important that the PCR amplification does not reach the plateau phase (Chapter 2) thus ensuring that the differential expression profile is mirrored in the corresponding amplified cDNA. 'Test' amplifications are required using different numbers of PCR cycles so that optimal conditions are used for the transcript in question. Following amplification, the cDNAs are size fractionated through an agarose gel and subjected to Southern blot analysis. *Figure 8.4* shows a comparison of a standard Northern blot using 2 µg of polyA RNA and a virtual Northern blot using 100 ng of total RNA. Virtual Northern blotting has been used successfully for a number of gene expression studies and it has been shown that as little as 100 cultured cells is sufficient to generate more than 100 virtual Northern blots (2).

Quantitative RT-PCR

Since every PCR displays different reaction dynamics it is difficult to compare semi-quantitative data from separate experiments, and comparisons of mRNA transcript levels from amplified genes using different primer pairs cannot be made. More robust and reliable methods for mRNA

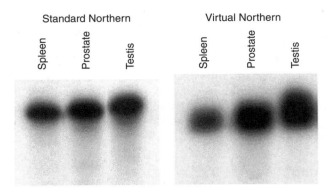

Figure 8.4

Comparison of a standard Northern blot analysis using 2 µg of polyA RNA and a virtual Northern blot using 100 ng of total RNA. (*Reproduced with permission of* CLONTECH *Laboratories Inc.*)

quantitation rely on the use of internal standards and quantitative competitive RT-PCR.

Competitor PCR

A relatively simple approach to quantitative RT-PCR involves coamplification of both the target mRNA and a standard RNA in a single reaction using primers common to both target and standard (*Figure 8.5*). As the standard competes with the target mRNA for both primers and enzyme it is referred to as a competitor or mimic (3). It is best to design an RNA competitor that is slightly different in length from the target allowing simple and direct gel determination of relative efficiencies of amplification. The competitor RNA can be generated by T7 or SP6 directed *in vitro* transcription from a suitable plasmid vector. The competitor should contain the same primer sites as the target and can then be used to control for both cDNA synthesis and PCR. Both the target and standard are primed with a gene-specific primer and the cDNAs are then coamplified directly in the same tube using a single primer pair. In practice several reactions are performed simultaneously with different amounts of competitor RNA. The concentration of the target mRNA can be determined as being equivalent to that of the competitor when there is a 1:1 ratio of target and competitor products. One of the most critical steps in this process is determining accurately the concentration of the competitor RNA. The best and simplest way of doing this is by spectrophotometry. The absorbance of the transcribed competitor RNA, after DNAse treatment, at 260 nm (A_{260}) should be measured in triplicate and the average will give a quantitatively accurate measure of the competitor RNA concentration.

Controls and measurements

In all experiments that involve the quantification of mRNA levels it is important to ensure the integrity of samples, and to ensure that normal-

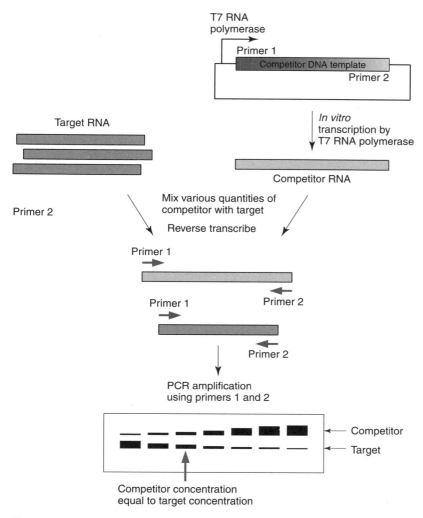

Figure 8.5

Principle of quantitative RT-PCR analysis using *in vitro* transcribed competitors. A competitor is generated that can be distinguished from the target product upon gel analysis. The RT-PCR reactions are spiked with known amounts of competitor. The concentration of competitor that gives the same amount of product as the target sample provides a measure of the amount of target mRNA in the original sample.

ization between samples can be achieved. This is done by including the analysis of a gene whose level should remain constant under all conditions. For example actin is widely used as such a control. The levels of the mRNA for this protein can be used to quantitate the amounts of mRNA produced from a sample, and differences in signal intensity can be used to moderate the levels of target gene signals. The measurement of signals from samples separated through gels will depend on whether the DNA is labeled or not.

For standard DNA gels, it is possible to capture gel images using a CCD camera and to analyze the intensity of the signals in each band by using appropriate software, often supplied by the manufacturer of the imaging equipment. These programs allow integration of the intensity of the band and provide a numerical value for the level of signal. The use of a standard, such as actin, allows the normalization of signal intensities. If the samples are radiolabeled, such as for virtual Northern analysis, then the signals can be measured by exposing X-ray film in a suitable cassette. It is important that the bands are gray and do not become black during this exposure since this prevents subsequent accurate quantification of signal intensities when the film is scanned in a densitometer. For faster and more accurate analysis use a phosphorimager, which has a much broader dynamic range than X-ray film. It uses a storage phosphor autoradiography system, but some instruments also offer direct fluorescence and chemifluorescence detection. All systems come with associated software for accurately quantifying signal intensities.

8.4 One-tube RT-PCR

RT-PCR protocols are not always successful. The major limitation is that cDNA synthesis is commonly performed at 42°C, which does not eliminate RNA secondary structures. In addition, the two-step procedure involving the first-strand cDNA synthesis step and then the PCR step can result in potential contamination problems. New systems have been developed where both the RT-PCR reaction and the subsequent PCR reaction are carried out in the same tube. Details of such systems are provided in Chapter 3. A further benefit is that in some systems cDNA synthesis can be performed at high temperatures, which eliminates RNA secondary structure. For example, the Titan™ one-tube RT-PCR system (Roche) uses a reverse transcriptase and buffer that allows the cDNA synthesis reaction to be performed at 60°C. The tube, now containing first-strand cDNA, can be directly subjected to PCR amplification, as the initial reactants include a thermostable DNA polymerase. This system has been used to successfully amplify cDNAs up to 6 kb in length from as little as 10 ng of total RNA.

8.5 Differential display

Differential display, first described by Liang and Pardee (4), allows rapid and simultaneous display of the expression profiles of mRNAs from different cell populations. The main steps include:

- reverse transcription using a 3'-anchored primer;
- PCR in the presence of α-^{35}S dATP using an arbitrary 5'-primer;
- size fractionation of the amplified products and comparison of patterns derived from different cell populations; and
- re-amplification and cloning of differentially expressed cDNA products.

Each step will be described in more detail, but for a comprehensive protocol see the website http://www.plant.dlo.nl/projects/hybtech/Liu/DISPLAY.html.

Reverse transcription

The 3'-primer for reverse transcription is based on the polyadenylation (polyA) tail found on eukaryotic mRNAs. An oligo-dT primer is used to anchor the primer at the 3'-end of the mRNA to ensure directional first-strand cDNA synthesis. If you tried to compare all the transcripts at one time the pattern would be extremely complicated and impossible to interpret. To simplify the interpretation the oligo-dT primer is modified to anneal to only a subset of mRNA molecules. At the 3'-end of the primer, one or commonly two extra bases are included to select a subpopulation of the mRNAs for amplification. This specificity of annealing shown below also ensures that all products prime from the 3'-end of the transcript rather than nonspecifically within the polyA tail:

```
mRNA  5'-NNNNNNNNNNNNNAAAAAAAAAAAAAAAAAAAAAAAAAAAAAAAAAAAAAAAAAA-3'
                      ||||||||||||||||||||||||||||||
                 3'-NNTTTTTTTTTTTTTTTTTTTTTTTTT-5'
```

NN in the primers could be either AA, AG, AC, GA, GG, GC, CA, CG, CC, AT, GT or CT, giving 12 different combinations of oligo-dT primer. Any single primer will therefore anneal to one-twelfth of the total mRNAs in the population. The use of all 12 primers in separate cDNA synthesis reactions should amplify different subpopulations of the mRNA complement of the cells thereby allowing comparison of essentially all the transcripts. The reverse transcription reaction is then performed as described previously (Section 8.2). Further advances in primer design were subsequently introduced. For example only a one-base anchor means that only three primers of the general design $N_{10}T_{11}C$ and $N_{10}T_{11}A$ and $N_{10}T_{11}G$, are required. In this case N_{10} represents a 10-nucleotide 5'-sequence that includes a restriction site for subsequent cloning (5). The two-nucleotide-anchor primers produce fewer bands per gel lane than the single-anchor primers, but provide higher resolution of the product bands (6).

```
mRNA  5'-NNNNNNNNNNNNNAAAAAAAAAAAAAAAAAAAAAAAAAAAAAAAAAAAAAAAAAA-3'
                      ||||||||||||||
                 3'-ATTTTTTTTTTTTGGAATTCCTA-5'
                 3'-GTTTTTTTTTTTTGGAATTCCTA-5'
                 3'-CTTTTTTTTTTTTGGAATTCCTA-5'
```

The PCR reaction

The main constraint in differential display is the separation and display of products. Amplified cDNAs larger than about 500 bp will not be resolved by standard polyacrylamide DNA sequencing gels. Thus it is important to try to amplify fragments from each cDNA within 500 bp of the mRNA polyA tail. This is most conveniently achieved using a short, essentially random sequence 5'-primer that is 10 nucleotides in length (4). A range of such primers is commercially available, for example from Operon Technologies. The specificity of amplification also increases dramatically if the final dNTP concentration is reduced to 2 µM compared with 200 µM used for standard PCR reactions. The lower dNTP concentration also increases the efficiency of incorporation of [α-^{35}S] dATP, increasing the specific activity of the generated fragments and consequently improving their detection. Differential display procedures normally require extensive optimization in order to efficiently and clearly display cDNA differences. Optimization is very important as the success of both the DNA elution and re-amplification depends on the amount of cDNA generated during the first round of PCR. A good way of optimizing the first round PCR is to take advantage of known genes that are differentially expressed in the cells that you will use for the 'real' experiment. The value of such an internal control was demonstrated with the murine thymidine kinase (TK) gene from tumorigenic cells (4).

Displaying the differentially expressed genes

Polyacrylamide gel electrophoresis can separate DNA molecules that differ by as little as 1 bp in 500 bp and is therefore an appropriate method for displaying differentially expressed genes. The gel system is the same as that used for manual DNA sequencing (7). The accuracy and resolution of the differential gene expression profiles depends to a large extent on the quality of the polyacrylamide gel and generally a final polyacrylamide concentration of 6% is appropriate with an effective separation range of between 25 and 500 bp. Acrylamide and bis-acrylamide are both neurotoxins which can enter the body by inhalation, if a powder, or through the skin, so extreme care should be taken when handling these chemicals, and protective clothing, gloves and mask should always be used. Because of this we recommend using commonly available ready prepared solutions.

Preparing the gel apparatus

A number of gel apparatus are commercially available and consist of two glass plates (a 'notched' front plate and a complete back plate), plastic spacers, a comb and a discontinuous electrophoresis buffer system. Generally, a gel of 40 cm length and 20 cm width is used. The gel thickness is determined by the spacer thickness and is normally between 0.2 mm and 0.6 mm. Thinner gels give increased resolution but are fragile, while thicker gels are easier to handle, accept larger sample volumes, but are more difficult to fix and dry. A gel thickness of 0.4 mm is recommended, which gives good resolution, ease in post-run handling and is generally easy to fix and dry.

Re-amplification and cloning

Once you are satisfied that there are cDNAs differentially expressed between your samples it is time to perform the re-amplification and cloning of the cDNA fragments. The re-amplification serves two purposes; first, it generates sufficient cDNA to clone into a plasmid for further analysis, and second, it serves as a control to demonstrate that the initial PCR amplification was primer-specific. To perform the re-amplification the DNA must be eluted from the dried gel by crushing the gel slice in elution buffer (http://www.plant.dlo.nl/projects/hybtech/Liu/DISPLAY.html).

The amount of cDNA available for re-amplification may be limiting and a frequent problem when analyzing the PCR re-amplification is that no product can be detected. This is not uncommon even after 40 rounds of

PCR and a third round of PCR amplification may be required. The problems associated with re-amplification, described in Chapter 4, such as increased probability of PCR-generated mutations, are not so important here since differential display cDNA fragments will usually be used to screen a cDNA library to isolate the full-length cDNA for further characterization. However, by increasing the number of PCR rounds, the possibility of amplifying nonspecific DNA fragments increases. This means that following cloning of the product, it is important to verify its differential expression character by Northern blot analysis or RT-PCR (Section 8.3).

Advantages of differential display

The advantages of PCR-based differential display are:

- that differences in expression patterns can be readily visualized by running different samples in parallel;
- the differentially expressed cDNAs can, in theory, be easily recovered, cloned and sequenced;
- the displayed mRNA patterns are highly reproducible.

In addition, PCR-based differential display is technically much simpler and quicker than the traditional techniques used for detecting differences in gene expression patterns. After initial optimization, PCR-based differential display only requires 2 days for differential band pattern visualization and only 5 days for the subsequent elution, re-amplification and cloning.

Fluorescent differential display

The main drawback to differential display is that it makes use of hazardous radioisotopes. Recent advances have overcome the use of radioisotopes and manual autoradiography detection by incorporating fluorescent primers or fluorescent dUTP into the PCR reaction as part of the differential display protocol. The fluorescent signal can then be detected using an automated fluorescent DNA gel imager (for example from Hitachi or Amersham Biosciences) and the fluorescent signal can be analyzed using various software packages such as FMBIO (Hitachi). This technique has been termed fluorescent differential display (FDD). The main advantages of FDD are: full automation, which makes it cost effective in terms of time when optimizing the experimental protocol; no need for radiolabels; high level of reliability; and consistency between duplicate samples. Conveniently, fluorescently labeled universal primers can be used for virtually all experiments, thereby reducing the overall cost. Both fluorescein isothiocyanate and rhodamine can be used as fluorescent tags. The FDD procedure is essentially identical to standard differential display.

FDD has been used successfully for a number of applications such as the identification of differentially expressed genes during neuroblastoma differentiation (8) and in identifying differentially expressed genes in plants in response to different light regimes (9). Recently it has been shown that cloning of the differentially expressed fragments can be omitted. After excision of FDD bands from the polyacrylamide gel these are separated on an agarose gel containing a base-specific DNA ligand which separates equally sized fragments differing in base composition. It has been shown that most of the cDNA fragments selected using this method can be directly sequenced and subsequent Northern blot analysis reveals them to have differential expression patterns (10).

8.6 PCR in a cell: *in situ* RT-PCR

The concept of performing PCR to detect gene expression patterns inside single cells or even specific intracellular organelles would have been unbelievable only a few years ago. *In situ* RT-PCR follows the same principle as RT-PCR but instead of being performed in a test-tube the reaction occurs inside cells or even at specific intracellular locations showing organelle-specific gene expression.

Principle

In situ RT-PCR detects gene expression profiles at the cellular and subcellular level. In essence the technique is carried out on the biological sample usually immobilized on a glass slide. *In situ* RT-PCR therefore has the sensitivity of standard RT-PCR but also the spatial resolution of *in situ* hybridization. The technique can be divided into several main steps including:

- sample preparation;
- *in situ* first-strand cDNA synthesis;
- *in situ* PCR using either labeled primers or unlabeled primers; and
- *in situ* hybridization or detection.

The following Sections outline these different steps whilst more detailed information can be found at http://www.bioscience.org/1996/v1/c/nuovo1/htmls/list.htm.

Sample preparation

Preparing a sample for *in situ* RT-PCR is slightly more time consuming than preparing samples for standard RT-PCR as they must maintain cellular integrity during first-strand cDNA synthesis, PCR and product detection. To achieve this, samples are normally prepared on glass slides to allow sufficient heat transfer during the RT and PCR steps. Depending on the tissue or sample there are a number of standard sample preparation steps that should be followed. To maintain cellular integrity the tissues should be fixed in 2–4% paraformaldehyde or 2–3% glutaraldehyde solutions. Once fixed the tissue must be sectioned, usually by embedding in paraffin-based waxes, followed by standard μm-range sectioning. The tissue sections are placed on silane-coated *in situ* PCR slides and allowed to dry at room temperature for several days. The sections are then deparaffinized, incubated in xylene and dried at room temperature.

The cells must be permeabilized to allow entry of reactants for the subsequent RT-PCR analysis which is normally achieved by pretreating sections with proteinases such as proteinase K or pepsin. It must be stressed that proteinase treatment should be optimized for each sample type, since over-digestion will result in loss of cellular integrity while under-digestion will render the sections impermeable to reaction components. A recommended starting point is incubation with 5 µg ml^{-1} proteinase K for 30 min at 37°C followed by heat inactivation at 95°C for 5 min.

As for standard RT-PCR, it is important to remove any contaminating DNA which can potentially interfere with the RT-PCR reaction. The removal of DNA for *in situ* RT-PCR is extremely important since product detection is not based on size determination, which eliminates the possibility of using intron-spanning primers. Sections should therefore be treated with RNAse-free DNAse. At this stage it is also important to include RNAse inhibitors to avoid RNA degradation. It has been shown that the precise adjustment of the DNAse concentration and incubation time is essential for reliable and reproducible results when performing *in situ* RT-PCR (1). The efficacy of DNAse treatment varies between different cell lines and it is more appropriate to fine-tune the incubation time rather than DNAse concentrations.

Attachment of samples to glass slides

As described above, in most cases *in situ* RT-PCR involves fixing samples to glass slides. It is important that samples remain attached to the glass slides throughout the entire experimental procedures. This requires the attachment procedure to be thermostable and chemically inert. The use of silane-coated slides for *in situ* PCR studies was first described by Dyanov and Dzitoeva (11). It allows rapid and irreversible sample attachment where more than 95% of the material remains attached after *in situ* PCR and *in situ* hybridization procedures. For most applications, silane A-174 (Bind-Silane; Amersham Pharmacia Biotech) together with γ-methacryloxypropyl-trimethoxy-silane (Sigma Chemicals) provides a very strong adhesive for a wide range of samples. For most single cell applications it is advisable to perform the fixation and permeabilization on the glass slides. Fixed and paraffin-embedded sections are generally easy to transfer to silane-coated glass slides. The attachment of whole organs to glass slides has obvious size limitations. Whole organs from, for example, *Drosophila melanogaster* have been successfully attached to glass slides and subjected to *in situ* RT-PCR experiments (11). The attachment process is essentially identical to the attachment procedure used for tissue sections.

In situ thermal cyclers

There are an increasing number of instruments available for *in situ* PCR applications and these are designed to accept multiple slides and to provide optimized thermal exchange for *in situ* applications (Chapter 3).

In situ first-strand cDNA synthesis

The *in situ* first-strand cDNA synthesis reaction is essentially identical to standard first-strand cDNA synthesis described in Section 8.2. One limitation is that the absolute level of mRNA is not controllable and this may, in cases of low mRNA levels, result in a low yield of first-strand cDNA molecules. This is normally not a problem and an example illustrating this is the successful detection of insulin-like growth factor-IA (*IGF-1A*) mRNA from human lung tumor cell lines (12). *IGF-1A* mRNA levels are present at extremely low levels in these cell lines (13), which limited the use of standard *in situ* hybridization protocols and Northern blot experiments. However, by carefully optimizing *in situ* RT-PCR protocols it was possible to detect *IGF-1A* mRNA species and detect their cellular location.

The *in situ* first-strand synthesis reaction can be performed using random primers, an oligo-dT primer, gene-specific primers, or a combination of these. Random primers will generate a vast array of differently sized single-stranded cDNA fragments which in turn will act as templates for the *in situ* PCR reaction. In most cases the randomly generated DNA fragments will span the region to be subsequently amplified; however it is possible that the majority of DNA fragments generated lie outside of the desired PCR amplification area. To ensure that the region of DNA to be subsequently amplified is present in the first-strand cDNA, it is recommended that either an oligo-dT or a gene-specific primer be used. As described in later parts of this Section, product detection can be achieved by indirect *in situ* hybridization or by using labeled primers for the *in situ* PCR. If *in situ* hybridization is the method of choice we advise the use of both oligo-dT and random primers to maximize the efficiency of the first-strand reaction. However, if labeled primers are to be used it is advisable to use gene-specific primers or an oligo-dT primer.

For the first-strand cDNA synthesis reaction a premix consisting of reaction buffer, dNTPs, RNAse inhibitor, the primer or primer set of choice at a concentration of 200 pmols and the reverse transcriptase. The premix is added to the sample on the in *situ* PCR slide and 'sealed' in a chamber. The chamber can be constructed with silicon spacers that surround the sample followed by sealing with a second glass slide. Alternately, specialized *in situ* glass slides, cover slips, and cover discs can be purchased from a number of commercial sources such as PE Biosystems and Hybaid. Once sealed the reaction should be allowed to proceed at 42°C for 1 h, or at a higher temperature if a thermostable reverse transcriptase is used.

In situ PCR

After first-strand cDNA synthesis the cover disc should be removed, the samples rinsed briefly in phosphate buffered saline (PBS) and the PCR premix added to the sample, and the chamber resealed. The PCR premix is essentially the same as for standard PCR containing the reaction buffer, dNTPs, gene-specific primers, and *Taq* DNA polymerase. Although the *in situ* PCR cycling conditions will not vary a great deal from standard PCR conditions some degree of optimization of reactant concentrations and amplification conditions may be required. For example, the number of amplification cycles required for *in situ* PCR is normally lower than for standard PCR. In general, 15–20 cycles are sufficient, with an increase in cycle number often having a detrimental outcome with the signal losing its 'crisp' appearance and becoming more diffuse due to excess final product (12).

Once the *in situ* PCR has been completed the reaction chamber should be rinsed thoroughly with PBS. There are two main ways in which to detect the *in situ* PCR-generated amplification products and these are described in the following Sections.

In situ hybridization

In situ hybridization is well established and has been optimized for a number of different biological systems. Some common applications include detection of gene expression patterns at the cellular level, cellular detection of pathogen DNA, detection of DNA rearrangements in single cells and analysis of both legitimate and illegitimate recombination events. The power of *in situ* hybridization makes it an obvious method for product detection as part of the *in situ* RT-PCR protocol.

The basic principle of *in situ* hybridization is the ability of a labeled nucleic acid fragment to 'seek out' and hybridize to a complementary nucleic acid sequence. The method is extremely versatile and by applying only slight modifications it can be used to detect either perfectly homologous nucleic acid sequences (high-stringency conditions) or heterologous stretches of nucleic acids (low-stringency conditions). When using *in situ* hybridization as part of the *in situ* RT-PCR protocol, high-stringency conditions are always required so that only DNA sequences generated as part of the *in situ* PCR are detected. High-stringency conditions will also minimize nonspecific background hybridization.

A variety of different labels, varying from radiolabels to enzymatic components, can be conjugated to probes, with enzyme-conjugated probes being the most common for *in situ* RT-PCR applications.

Alkaline phosphatase

There are several different commercially available enzyme-linked labels that can be incorporated into nucleic acids. Alkaline phosphatase (AP) is widely used for *in situ* RT-PCR applications and several chromogenic substrates such as 5-bromo,4-chloro,3-indolyl phosphate (BCIP) and nitro blue tetrazolium (NBT) are cleaved by the enzyme to generate a visible dense blue insoluble precipitate. Both BCIP and NBT are stable as stock solutions. AP can be linked to DNA fragments in different ways. A common and efficient method is to conjugate avidin-bound biotinylated AP to the generated probe, and several manufacturers offer kits for such enzyme–DNA probe conjugation. Enzyme-linked probes can be stored for extended periods at 4°C. The DNA probes for *in situ* hybridization reactions are normally generated by PCR amplification or restriction digestion of a plasmid followed by gel purification as described in Chapter 6.

Once the probe has been labeled the *in situ* hybridization reactions can be performed. It is important to prehybridize the sample to minimize nonspecific hybridization. Prehybridization makes use of a blocking agent such as sonicated salmon sperm DNA or calf thymus DNA to 'mask' nonspecific targets for the probe. The prehybridization buffer should be of the same composition as the hybridization buffer with the exception of the added probe. A 'standard' hybridization buffer, for an enzyme-conjugated probe, consists of $1 \times$ PBS and sonicated 'blocking' DNA at a final concentration of 200 µg ml^{-1}. Samples should be incubated in prehybridization buffer at 37–40°C for 2–5 hours with gentle shaking followed by careful rinsing in hybridization buffer and addition of fresh hybridization buffer containing the denatured labeled probe. The labeled probe must be added to the hybridization buffer (2–5 µg ml^{-1}) before adding it to the sample to eliminate high probe concentrations at the site of application. The hybridization reaction should be performed at the same temperature as the prehybridization for between 8 and 16 hours. However, to limit potential loss of cellular integrity and minimize loss of enzyme activity the incubation times should be kept to a minimum.

Post-hybridization washes are important to remove nonspecifically bound probe molecules. The substrates NBT or BCIP can then be added in the reaction buffer. AP requires a basic pH for optimal activity and the buffer contains 100 mM NaCl, 50 mM MgCl$_2$, and 100 mM Tris, pH 9.5. The reaction should be performed in the dark for between 10 and 15 minutes or until visible staining appears. The reaction can be stopped by extensive washing in PBS containing 20 mM EDTA followed by viewing using bright-field microscopy.

Indirect detection of incorporated labels

Another well-documented method of product detection as part of *in situ* RT-PCR is the use of modified nucleotides followed by antibody detection. For *in situ* RT-PCR applications, as for many other applications, digoxigenin-11-2′-deoxyuridine-5′triphosphate (DIG-dUTP) is commonly used as the modified nucleotide. Digoxigenin is a steroid hapten found only in *Digitalis* plants so background problems are minimal. DIG-dUTP is incorporated into DNA as part of the PCR (Chapter 3) and the amplification products can be analyzed by agarose gel electrophoresis (Chapter 5) and gel purified (Chapter 6) for use as a probe (*Figure 8.6*). The DIG-labeled probe can be used for the hybridization steps as described above, but normally at a higher temperature such as 65°C. After hybridization and post-hybridization washes the samples should be incubated at room temperature with an AP-conjugated anti-DIG antibody for 1 h. After extensive washing with PBS, product detection is performed using NBT or BCIP. One limitation when using anti-DIG antibodies is background staining in complex tissues due to antibody 'trapping'. However, by including appropriate controls, false staining patterns can normally be identified.

Labeled primers and direct detection

Recent developments have eliminated the need for post-PCR *in situ* hybridization. Direct detection of the *in situ* RT-PCR products can be achieved using primers carrying a fluorescent label, usually fluorescein, in the final PCR reaction (14). An example of the use of fluorescein-labeled primers for final product detection is illustrated by detection of tumor necrosis factor (TNF). The mammary carcinoma cell line MCF-7 harbors the *TNF-α* gene but does not express *TNF* mRNA. To achieve *TNF* expression the cell line has to be transduced with a retrovirus containing the cDNA for the human *TNF* gene. This system was chosen by Stein *et al.* (1) to develop a one-step *in situ* RT-PCR protocol. To check for successful *TNF*

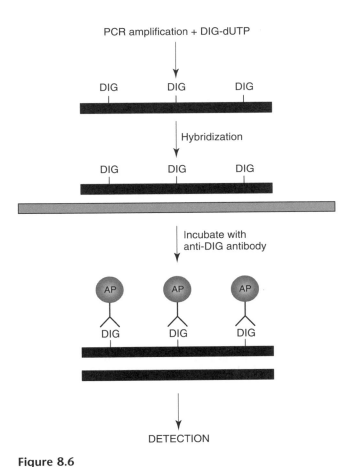

PCR amplification + DIG-dUTP

DIG DIG DIG

Hybridization

DIG DIG DIG

Incubate with
anti-DIG antibody

AP AP AP

DIG DIG DIG

DETECTION

Figure 8.6

Labeling of a PCR-amplified probe with DIG as part of *in situ* RT-PCR indirect
detection of gene expression.

expression standard RT-PCR was performed showing specific amplification
of the *TNF* gene. For *in situ* RT-PCR the transduced cell line was grown,
fixed and DNAse treated. M-MLV reverse transcriptase was used together
with random hexamer primers for the *in situ* RT reaction. For *in situ* PCR a
3′-unlabeled *TNF*-specific primer was used together with a 5′-fluorescein-
labeled TNF-specific primer, both at 1 µM concentration. The PCR was
performed *in situ* using a GeneAmp® In Situ System 1000 thermal cycler (PE
Biosystems). Slides were washed in PBS and the final amplification product
visualized by fluorescent microscopy (absorbance wavelength 495 nm and
emission wavelength 525 nm). This technique proved to be reliable and
reproducible and did not generate any 'false' positive results due to nonspecific amplification as shown when the nontransduced cell line was used
as a negative control. Also, as for any PCR, single primer controls, no primer
controls and no DNA polymerase controls were performed confirming
gene-specific amplification. This elegant technique can be modified by

using different fluorescent labels for different primer sets, which should make it possible to detect several expressed genes in one reaction.

As labeled primers can be expensive it is advisable to first produce an unlabeled primer set which should be used for the optimization of *in situ* PCR conditions. One of these primers can subsequently be used as the unlabeled primer for the 'real' experiment. The cost of having one unlabeled primer is minimal if it ensures that the expensive labeled primer will function efficiently. Optimization of conditions, such as annealing temperature and extension times, can be performed on genomic DNA or on *in vitro* generated first-strand cDNA, which makes the procedure rapid and easy to perform. Once optimized, the conditions can be transferred to the *in situ* RT-PCR protocol.

8.7 Microarrays

With the increasing availability of genome sequence information it has now become possible to interrogate the gene expression patterns in particular cells. To investigate which genes are expressed in particular cell types, or which genes are altered in their expression in response to some external stimulus or disease state, it would be useful to have a method for global analysis of gene expression. A DNA microarray represents an array of DNA sequences representing large numbers of genes, immobilized on a solid support, such as a nylon membrane, glass slide or silicon chip. The array must be prepared so that the identity of the sequence at each spot is known. The DNA samples spotted onto the support can be either cDNA sequences, which can be produced by PCR, or oligonucleotides. In the case of Affymetrix gene chips the oligonucleotides are actually synthesized *in situ* on the slide at defined locations.

The microarray chip is then interrogated by measuring the ability of mRNA molecules to bind or hybridize to the DNA template that produced it. Since the array contains many DNA sequences it is possible to determine from a single experiment the expression levels of hundreds or thousands of genes within a sample by measuring the amount of mRNA bound to each spot (or sequence) on the array. Usually the experiment is performed as a competition hybridization between two samples of mRNA, one from a control sample and one from the test sample. These mRNAs are used to produce cDNA populations that are differently labeled by addition of a fluorescent tag, say green for the control and red for the test sample. If there is more of a particular sequence in the control sample, the spot will be green, indicating that expression of the gene was reduced in the test sample. This is because more of the sequences that are available for hybridization to the chip sequence will be labeled green, and so more of these sequences will hybridize compared with the lower concentration red sequences. Conversely if the gene shows increased expression in the test sample the spot will be red, while if there is no difference in expression between the two samples the spot will be yellow indicating roughly equal amounts of both sequences have hybridized to the chip.

After the hybridization process the microarray slide is placed in a scanner that contains a laser to excite the fluorescent tags on the cDNAs; the resulting fluorescence is detected by a microscope and camera that captures an image of the array. The data are saved in the computer and analyzed by software programs which subtract the background fluorescence level from each spot and then calculate the relative intensity of green versus red. This gives a measure of the relative levels of the competing mRNAs bound to each spot and so provides information on the relative level of expression of the genes on the array. *Figure 8.7* shows a simple schematic of a microarray experiment. There are further details of microarray experiments at the NCBI website http://www.ncbi.nlm.nih.gov/About/primer/microarrays.html. There is also a simple animation that outlines the processes involved in a microarray experiment at http://www.bio.davidson.edu/courses/genomics/chip/chip.html.

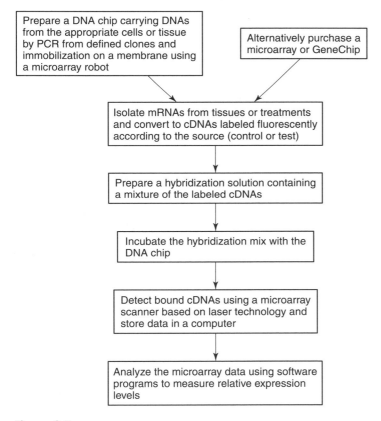

Figure 8.7

Scheme showing the basic steps in a microarray experiment.

8.8   RNA interference (RNAi)

RNA interference (RNAi) is an important approach for studying the importance of expression of specific genes within cell, tissues or organisms. It uses double-stranded RNA to interfere with gene expression and was first demonstrated in the free-living nematode *Caenorhabditis elegans* (15). It is now known to be a phenomenon of wide occurrence in many organisms including mammals (16–18), plants (19,20), insects (21), protozoans (22) and fungi (23). It provides a potentially powerful investigative tool for the genome-wide identification of gene function (24).

The basic mechanism of RNAi involves introducing a double-stranded RNA (dsRNA) of around 300–500 bp into a cell by microinjection or feeding. This dsRNA molecule is acted upon by an RNAse III-like enzyme called DICER which results in a processive ATP-dependent cleavage to generate a series of 21–23 bp duplexes that have 2 nt overhangs at their 3′-ends. These small interfering RNAs (siRNAs) can in some organisms be transmitted systemically to other cells through the involvement of the SID protein, a transmembrane channel protein. The siRNA forms a complex with a further nuclease called RNA-induced silencing complex (RISC). Once again, in an ATP-dependent process the siRNA duplex is unwound into single strands, thus activating RISC which then identifies homologous mRNA transcripts by base pairing with the single-strand siRNA and cleaving the mRNA and leading to mRNA degradation.

In order to generate the dsRNA it is first necessary to start with cDNA clones. These could be ESTs (Chapter 10) or clones isolated from a custom-made library. The usual approach is to use PCR to amplify a fragment of ~500 bp and to clone this into a vector containing two T7 promoters such as pPD129.36. By restricting aliquots of the recombinant vector adjacent to one or other of the T7 promoters it is possible to create linear templates for production of either the sense or the antisense RNA strand of the sequence by using T7 RNA polymerase. These RNA samples are then mixed and annealed to give double-stranded RNA that can be introduced into the test cells or organisms. For mammalian cell culture experiments it is now more common to use siRNAs that are prepared synthetically to induce an RNAi effect, as these appear to be more efficient in such systems and are more readily taken up by the cells. For many organisms, however, siRNAs do not mediate an RNAi effect and so longer dsRNAs are essential.

In most RNAi experiments one measures a phenotype mediated by the dsRNA compared with a treatment with a control RNA from a nontarget gene. This allows one to discount any nonspecific effects due to treatment with any dsRNA, and to identify those specifically mediated by the test sequence. In some cases, however, and particularly in mammalian cell culture experiments, it can be useful to measure changes in gene expression mediated by the treatment. The more sensitive and appropriate approach to such analysis is real-time PCR, described in detail in Chapter 9.

References

1. Stein U, Walther W, Wendt J, Schild TA (1997) *In situ* RT-PCR using fluorescence-labeled primers. *Biotechniques* **23**: 194–195.
2. Franz O, Bruchhaus I, Roeder T (1999) Verification of differential gene transcription using virtual northern blotting. *Nucleic Acids Res* **27**: e3.
3. Siebert PD, Kellogg DE (1995) PCR MIMICs: competitive DNA fragments for use in quantitative PCR. In McPherson MJ, Hames BD, Taylor GR (eds) *PCR 2: A Practical Approach*, pp. 135–148. Oxford University Press, Oxford, UK.
4. Liang P, Pardee AB (1992) Differential display of eukaryotic messenger RNA by means of the polymerase chain reaction. *Science* **257**: 967–971.

5. Liang P, Zhu W, Zhang X, Guo Z, O'Connell RP, Averboukh L, Wang F, Pardee AB (1994) Differential display using one-base anchored oligo-dT primers. *Nucleic Acids Res* **22**: 5763–5764.
6. Martin KJ, Pardee AB (1999) Principles of differential display. *Methods Enzymol* **303**: 234–257.
7. Brown TA (1994) *DNA Sequencing: The Basics*. BIOS Scientific Publishers, Oxford, UK.
8. Kito K, Ito T, Sakaki Y (1997) Fluorescent differential display analysis of gene expression in differentiating neuroblastoma cells. *Gene* **184**: 73–81.
9. Kuno N, Muramatsu T, Hamazato F, Furuya M (2000) Identification by large-scale screening of phytochrome-regulated genes in etiolated seedlings of *Arabidopsis* using a fluorescent differential display technique. *Plant Physiol* **122**: 15–24.
10. Yoshikawa Y, Mukai H, Asada K, Hino F, Kato I (1998) Differential display with carboxy-X-rhodamine-labeled primers and the selection of differentially amplified cDNA fragments without cloning. *Anal Biochem* **256**: 82–91.
11. Dyanov HM, Dzitoeva SG (1995) Method for attachment of microscopic preparations on glass for *in situ* hybridization, PRINS and *in situ* PCR studies. *Biotechniques* **18**: 822–824.
12. Martinez A, Miller MJ, Quinn K, Unsworth EJ, Ebina M, Cuttitta F (1995) Non-radioactive localization of nucleic acids by direct in situ PCR and in situ RT-PCR in paraffin-embedded sections. *J Histochem Cytochem* **43**: 739–747.
13. Reeve JG, Brinkman A, Hughes S, Mitchell J, Schwander J, Bleehen NM (1992) Expression of insulinlike growth factor (IGF) and IGF-binding protein genes in human lung tumor cell lines. *J Natl Cancer Inst* **84**: 628–634.
14. Long AA, Komminoth P, Lee E, Wolfe HJ (1993) Comparison of indirect and direct in-situ polymerase chain reaction in cell preparations and tissue sections. Detection of viral DNA, gene rearrangements and chromosomal translocations. *Histochemistry* **99**: 151–162.
15. Fire A, Xu SQ, Montgomery MK, Kostas SA, Driver SE, Mello CC (1998) Potent and specific genetic interference by double-stranded RNA in *Caenorhabditis elegans*. *Nature* **391**: 806–811.
16. Zamore PD, Tuschl T, Sharp PA, Bartel DP (2000) RNAi: double-stranded RNA directs the ATP-dependent cleavage of mRNA at 21 to 23 nucleotide intervals. *Cell* **101**: 25–33.
17. Hannon GJ (2002) RNA interference. *Nature* **418**: 244–251.
18. Silva J, Chang K, Hannon GJ, Rivas FV (2004) RNA-interference-based functional genomics in mammalian cells: reverse genetics coming of age. *Oncogene* **23**: 8401–8409.
19. Waterhouse PM, Graham MW, Wang MB (1998) Virus resistance and gene silencing in plants can be induced by simultaneous expression of sense and antisense RNA. *Proc Natl Acad Sci USA* **95**: 13959–13964.
20. Jorgensen RA, Cluster PD, English J, Que Q, Napoli CA (1996) Chalcone synthase cosuppression phenotypes in petunia flowers: comparison of sense vs antisense constructs and single-copy vs complex T-DNA sequences. *Plant Mol Biol* **31**: 957–973.
21. Kennerdell JR, Carthew RW (1998) Use of dsRNA-mediated genetic interference to demonstrate that *frizzled* and *frizzled 2* act in the wingless pathway. *Cell* **95**: 1017–1026.
22. Ngo H, Tschudi C, Gull K, Ullu E (1998) Double-stranded RNA induces mRNA degradation in Trypanosoma brucei. *Proc Natl Acad Sci USA* **95**: 14687–14692.
23. Cottrell TR, Doering TL (2003) Silence of the strands: RNA interference in eukaryotic pathogens. *Trends Microbiol* **11**: 37–43.
24. Kamath RS, Ahringer J (2003) Genome wide RNAi screening in *Caenorhabditis elegans*. *Methods Enzymol* **30**: 313–321.

Protocol 8.1 Reverse transcriptase reaction

EQUIPMENT

Adjustable heating block or water bath

Gel electrophoresis tank

MATERIALS AND REAGENTS

RNA isolation kit (e.g. Qiagen RNeasy® minikit)

First-strand cDNA synthesis kit (e.g. ProSTAR™ First-strand RT-PCR Kit, Stratagene)

or

Reverse transcriptase buffer: (AMV 10 × RT buffer; 25 mM Tris-HCl pH 8.3, 50 mM KCl, 2 mM DTT, 5 mM MgCl$_2$)

10 mM dNTP mix

Random hexamer, gene-specific or oligo-dT primer

RNase inhibitor

Reverse transcriptase (e.g. Qiagen)

0.8% agarose (100 ml; 0.8 g of agarose in 100 ml of 1 × TAE)

1. Add to a reaction tube the following in order:
   - 1–5 µg of RNA (the amount of RNA used for the RT reaction depends on the abundance of the target transcript);
   - 1 mM of each dNTP;
   - 50–100 pmol of primer (random hexamers, gene-specific primers, or oligo-dT primers);
   - 1–5 units of an RNase inhibitor;
   - water up to 49 µl.

2. Heat the reactions to 90°C for 5 min. (Heating the RNA eliminates RNA secondary structures and in turn increases the priming efficiency.)

3. Add 50–200 units of reverse transcriptase.

4. Incubate reactions at 42°C for 1 h.

5. Heat inactivate the reaction at 90°C for 10 min. (Heating the reaction to 90°C inactivates the reverse transcriptase and denatures remaining RNA–DNA hybrids.)

Real-time RT-PCR

9

9.1 Introduction

The importance of mRNA (transcript) quantification and profiling in basic research, molecular diagnosis and biotechnology has led to a rapid advance in quantitative reverse transcriptase PCR (RT-PCR) technologies in terms of instrumentation, automation and chemistries. Real-time RT-PCR represents an advance which has several advantages over conventional quantitative RT-PCR. The main advantage of real-time RT-PCR is that it sensitively and reproducibly quantifies the initial amount of starting template (transcript) by monitoring PCR amplification product (amplicon) accumulation during each PCR cycle, in contrast to conventional methods which detect the final end product. Furthermore, real-time RT-PCR is rapid, it is possible to analyze several transcripts (genes) simultaneously and post-PCR quantification procedures are eliminated. Because post-PCR analysis is eliminated carryover contamination is reduced and a higher throughput can be achieved. In addition, the dynamic range of real-time RT-PCR is higher (up to 10^{10}-fold) than conventional quantitative RT-PCR (1000-fold), which means that a wide range of amplification products can be accurately and reproducibly quantified. Although real-time RT-PCR does have numerous advantages over conventional methods there are also some disadvantages: higher costs and the inability to detect amplicon size can make differentiation between cDNA and DNA amplification difficult.

This Chapter deals with various aspects of real-time RT-PCR including detection systems and chemistries, oligonucleotide primer and probe design, real-time thermal cyclers, quantification and control selection, common pitfalls and applications. There is also a glossary of frequently used terms (*Table 9.1*) which will help you to familiarize yourself with the jargon of real-time RT-PCR.

9.2 Basic principles of real-time RT-PCR

The basic principle of real-time RT-PCR is much like conventional RT-PCR: cDNA is synthesized from mRNA using reverse transcriptase followed by cDNA PCR amplification and amplicon quantification. However, there are two fundamental differences: (i) amplicon accumulation is detected and quantified using a fluorescent reporter and not by conventional gel electrophoresis (Chapter 5); and (ii) amplicon accumulation is measured during each PCR cycle in contrast to standard end-point detection (Chapters 1 and 5). In addition, real-time RT-PCR is performed in 96-well microtiter plates and the fluorescent signal (amplicon accumulation) is detected and quantified using a real-time PCR thermocycler (Section 9.5).

A real-time RT-PCR reaction contains all the components used for conventional RT-PCR (Chapter 8) but in addition contains a fluorescent

Table 9.1 Glossary of frequently used terms

Term	Explanation
Amplification plot	A plot showing the cycle number versus the fluorescent signal which correlates with the initial amount of RNA during the exponential phase of the PCR amplification
Baseline	The initial cycles of PCR when the fluorescent signal shows no or little change
C_T value/number	The threshold cycle (C_T) indicates the cycle number where the reaction fluorescence crosses the threshold. This reflects the point in the reaction when sufficient amplicons have been generated to give a significant fluorescent signal over the baseline
Linear dynamic range	Range of initial template concentration giving accurate C_T values
Melting curve analysis	The melting point (T_m) of double-stranded DNA is the temperature at which 50% of the DNA is single-stranded and this temperature depends on the DNA length and GC content. When using SYBR® Green I a sudden decrease in fluorescence is detected when T_m is reached (dissociation of DNA strands and release of SYBR® Green I)
Multiplex analysis	Multiple RNA target analysis within one reaction well using gene-specific probes containing different reporter fluorophores
Passive reference	An internal reference dye to which the reporter dye can be normalized during analysis to correct for fluctuations between reaction wells. The most commonly used reference dye is ROX
Rn value	The normalized reporter (Rn) value represents the reporter dye fluorescence emission intensity divided by the passive reference dye. The Rn+ value is the Rn of a complete reaction sample whilst the Rn– value is the Rn of an unreacted sample. Rn– values are obtained during early amplification cycles (baseline)
ΔRn	(Rn+) – (Rn–) = the fluorescent signal intensity generated at each time point during PCR given a certain set of conditions
Standard curve	A curve consisting of C_T values plotted against the log of standard concentrations. The concentration/quantity of unknown samples are extrapolated from the standard curve

reporter either in the form of a fluorescent DNA-binding dye or as a fluorescent oligonucleotide primer (Section 9.3). Because the fluorescent reporter only fluoresces when associated with the product amplicon (Section 9.3), the increase in recorded fluorescence signal during amplification is in direct proportion to the amount of amplification product in the reaction. So how can this be related back to the initial starting amount of nucleic acid template in your sample? Because the intensity of the fluorescence emission is monitored and recorded during each PCR cycle it is possible to identify the exact PCR cycle at which the fluorescent signal significantly increases, and this correlates to the initial starting amount of the template. One easy way of thinking about this is that the higher the template concentration the earlier a significant increase in fluorescence will be observed. So what is a significant increase in fluorescence? During the early stages of the amplification reaction (3–15 cycles as a general rule) the fluorescence signal will show no or little change and this is taken as the background fluorescence or baseline (*Table 9.1*). By setting a fixed fluorescence threshold value above the baseline a significant increase in fluorescence is recorded when the signal intensity is higher than the threshold value, which in turn determines the threshold cycle or C_T (*Table 9.1*; *Figure 9.1(A)*). By knowing the C_T value for a reaction and by generating a standard cDNA concentration curve (fluorescence vs cDNA concentration) the concentration of the initial starting template can be extrapolated (*Figure 9.1(B)*). Although the overall principle of real-time RT-PCR can sound a little complicated the technique is relatively simple because of recent improvements in chemistries and instrumentation.

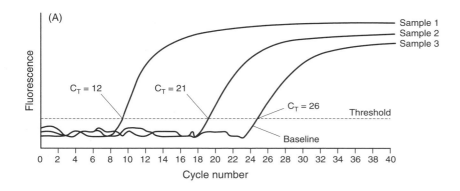

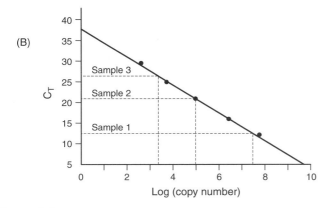

Figure 9.1

A schematic diagram showing a typical real-time RT-PCR reaction data set from three unknown samples. (A) The measured fluorescence signal at each amplification cycle is plotted against cycle number. The baseline where there is no or little change in fluorescence is indicated, as is the threshold value. The C_T value for each sample represents the cycle number when the fluorescence signal is higher than the threshold value. (B) A standard curve showing C_T versus the log of the copy number of standard cDNA samples. The initial starting concentration (copy number) of samples 1, 2 and 3 can be determined based on their C_T values. From this data set it is clear that sample 1 contains more starting template than sample 2 and that sample 3 has the least amount of starting template.

9.3   Detection methods

During the last 5 years there have been several advances in real-time RT-PCR detection methods in terms of both instrumentation and chemistries. Two general amplicon-detection methods are used which are based on either fluorescent DNA-binding dyes or fluorescent probes. The different detection systems, and their advantages and disadvantages, will now be described in detail.

SYBR® Green I

SYBR® Green I is a fluorescent DNA intercalating agent that binds to the minor groove of double-stranded DNA and upon excitation (498 nm) emits light (522 nm) that can be recorded by real-time PCR thermocyclers (Section 9.5). Because SYBR® Green I does not bind to single-stranded DNA and because the dye only emits weak fluorescence in solution, SYBR® Green I is widely used as a fluorescent reporter in real-time RT-PCR experiments to monitor double-stranded amplicon production (*Figure 9.2*). SYBR® Green I has numerous advantages over fluorescent probe approaches (covered in the remainder of this Section). First, SYBR® Green I is a nonsequence-specific dye which means that it will bind to any double-stranded piece of DNA. The advantage of this feature is that SYBR® Green I can be used for the amplification and monitoring of any gene. Second, because of the

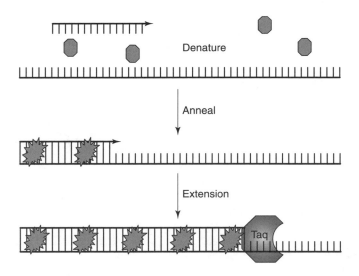

Figure 9.2

A schematic diagram showing how SYBR® Green I acts as a double-stranded-specific fluorescent reporter during PCR amplification. During the denaturation step when the DNA is single-stranded SYBR® Green I is free in solution in a nonfluorescent state (●). Upon annealing of the primer to the target template and during the extension phase the nonfluorescent SYBR® Green I (●) binds to the double-stranded amplicon and becomes fluorescent (✱).

nonsequence-specific nature of SYBR® Green I it represents a cheap alternative to fluorescent probes. Third, SYBR® Green I is simple to use. Fourth, SYBR® Green I is temperature stable and does not interfere with DNA polymerase.

Although the advantages of SYBR® Green I are clear there are also a few disadvantages compared with the fluorescent probe approach. Because SYBR® Green I binds to any double-stranded piece of DNA it will also bind to primer-dimers and any nonspecific amplification product. However, the nonspecific binding to primer-dimers can be overcome by melting curve analysis (*Table 9.1*) which will determine at which temperature primer-dimers are denatured (and therefore will stop fluorescing), allowing the identification of the target amplicon at an appropriate higher temperature. Melting curve analysis can also be used to eliminate the detection of nonspecific amplification products. Because of these potential problems SYBR® Green I-based real-time RT-PCR does need optimization and sometimes independent verification. Another potential problem that may occur is related to amplicon size. Long amplicons can generate a very strong fluorescence signal which may in turn saturate the camera situated inside the real-time PCR thermocycler. However, this is a minor problem since the size of the amplicon can easily be controlled by designing oligonucleotide primers that will amplify a 200–300 bp amplicon. SYBR® Green I is also most frequently used for single-amplicon monitoring (singleplex reaction) because of its nonsequence specificity, although multiplex reactions are possible if it is combined with melting curve analysis.

General principle of fluorescent probes

In contrast to fluorescent DNA-binding dyes such as SYBR® Green I, fluorescent probe approaches are based on amplicon detection using DNA sequence-specific oligonucleotide probes. These probes contain both a fluorogenic dye and a quencher dye and are designed to hybridize to the target gene either in between the two oligonucleotide primers used for the PCR amplification (TaqMan and Molecular Beacons; see below) or as part of one of the oligonucleotide primers used for the amplification reaction (Scorpions; see below). The general principle of amplicon detection using such probes is based on the fact that if a fluorescent dye is in close proximity to a quencher dye the fluorescent signal generated by the fluorescent dye in response to excitation is 'absorbed' by the nearby quenching dye, resulting in no fluorescent signal. This phenomenon is termed fluorescence resonance energy transfer (FRET). However, upon PCR amplification the fluorescent dye and the quenching dye become spatially separated either by probe displacement (TaqMan) or by probe rearrangement (Molecular Beacons and Scorpions), resulting in loss of FRET between the fluorescent dye and quenching dye, ultimately producing a fluorescent signal.

There are two main advantages of fluorescent probes over fluorescent DNA-binding dyes. First, because of the sequence specificity between the fluorescent probe and the target gene, detection of nonspecific amplification products and primer-dimers is eliminated. This reduces the need for extensive PCR optimization as is the case when using fluorescent DNA-binding dyes. Second, multiple probes containing different fluorogenic dyes emitting different wavelengths of light can be used in a single reaction, allowing for the detection of several amplicons simultaneously (multiplexing). As with any system, fluorescent probes also have disadvantages. Firstly, the coupling of a fluorescent dye and a quenching dye to an oligonucleotide can be costly, and secondly, the designed oligonucleotide probe can only be used for a single target gene. If the transcript levels of several genes are to be analyzed the use of fluorescent probes can soon become a costly exercise.

TaqMan probes

The TaqMan real-time RT-PCR assay was first reported in 1996 in two articles published by Williams' research group at Genentech in California (1,2) and is used widely in both basic and applied research programmes. The TaqMan assay combines the fact that *Taq* DNA polymerase has $5' \rightarrow 3'$ exonuclease activity and that dual-labeled oligonucleotide probes only fluoresce when cleaved/degraded by this exonuclease activity. In a typical TaqMan reaction three oligonucleotides are included: one forward primer, one reverse primer and one nonextendable internal TaqMan probe (*Figure 9.3*). The TaqMan probe is a standard oligonucleotide which has a covalently attached fluorescent reporter dye, such as FAM (6-carboxyfluorescein) at its 5'-end and a quencher dye, such as TAMRA (6-carboxytetramethylrhodamine) at its 3'-end (*Figure 9.3*). In addition to the most commonly used FAM and TAMRA dyes, 4,7,2',4',5',7'-hexachloro-6-carboxyfluorescein (HEX) and 4,7,2',7'-tetrachloro-6-carboxyfluorescein (TET) can be used as fluorescent dyes together with rhodamine or DABCYL as the 3'-quencher. When the TaqMan probe is intact, either free in solution or hybridized to its target DNA, the reporter dye fluorescence is absorbed by the quencher dye, because their close proximity allows FRET to occur (*Figure 9.3*). However, as the PCR reaction proceeds, the *Taq* polymerase will reach the 5'-end of the TaqMan probe and will strand displace it from the template. The $5' \rightarrow 3'$ exonuclease activity of the *Taq* polymerase will cleave the 5'-FAM dye from the probe thereby liberating the fluorescent reporter from its association with the quencher dye which leads to an increase in fluorescence (*Figure 9.3*). The increase in fluorescence is measured during each cycle and is proportional to the rate of probe displacement and hence the amount of amplification product in the reaction.

Although the TaqMan assay uses universal thermal cycling parameters and PCR reaction conditions, care should be taken when designing a TaqMan oligonucleotide probe. TaqMan probes should generally be longer than the amplification primers and typically between 20 and 30 nucleotides. In addition, the melting temperature (T_m) of a TaqMan oligonucleotide should be approximately 10°C higher than for the amplification primers, which allows hybridization to the target gene during the extension step. This is critical to ensure that the emitted fluorescence after TaqMan probe displacement is directly proportional to the amount of target DNA present in the reaction. Another critical factor when designing TaqMan probes is to avoid guanosine at the 5'-end

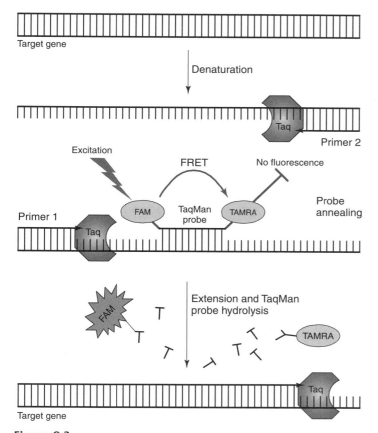

Target gene

Denaturation

Taq

Primer 2

Excitation

FRET   No fluorescence

FAM   TaqMan   TAMRA   Probe
       probe   annealing

Primer 1

Taq

Extension and TaqMan
probe hydrolysis

FAM

T

T   T   TAMRA
    T   T T

Target gene

Figure 9.3

A schematic diagram showing the principle of real-time RT-PCR using the TaqMan approach. In addition to general PCR components the reaction also contains a target gene-specific probe containing a fluorescent dye (FAM) at the 5′-end and a quencher dye (TAMRA) at the 3′-end. During the annealing step both the primers and the probe anneal to the target gene and because the quencher dye is in close proximity to the fluorescent reporter dye (on the same oligonucleotide) no fluorescence is generated. During the extension step the 5′→3′ exonuclease activity of *Taq* DNA polymerase displaces (degrades) the TaqMan probe resulting in loss of quenching and a fluorescence signal is generated ( ).

because this results in quenching of the fluorescent signal even after probe cleavage. Furthermore, the probe should contain more cytosine than guanosine and this can be achieved by designing either a sense or an antisense probe. Another critical factor when designing TaqMan probes is the possibility of coamplification of genomic DNA together with the target cDNA. To avoid this the probe should be designed so that it spans two exons, so that only correctly spliced variants of the cDNA are amplified. However, if the DNA sequence of the target cDNA has not been determined the only way to avoid genomic DNA amplification is to treat the RNA with RNAse-free DNAse.

Once the TaqMan probe has been designed, following the simple guidelines described above, little optimization is needed. Despite this, TaqMan probes can be expensive to synthesize and a separate probe is needed for each target gene. In addition, the TaqMan approach is not as sensitive as other more recent approaches such as molecular beacons (see below) and often has high background fluorescence.

Molecular beacons

As for TaqMan probes, molecular beacons contain a fluorescent and a quenching dye. Although molecular beacons make use of the fact that FRET occurs between a fluorescent and quenching dye when in close proximity, their design varies from that of TaqMan probes. During the annealing step the molecular beacon hybridizes to the target DNA, thereby separating the fluorescent reporter and the quenching dye, resulting in loss of FRET and an increase in fluorescence. During the PCR, molecular beacons remain intact and rehybridize during each cycle and because of this the fluorescence emission after hybridization is proportional to the concentration of target DNA (*Figure 9.4*).

Molecular beacons contain two parts: the probe, which can specifically hybridize to the target DNA; and the stem, which forms a hairpin structure whilst free in solution, ensuring that the fluorescent dye and the quenching dye are in close proximity, allowing FRET to occur (*Figure 9.4*). The first consideration when designing a molecular beacon is the selection of the probe sequence. The probe sequence can be any sequence within the amplicon that lies between the two oligonucleotide primers used for the amplification. The probe sequence should be between 15 and 30 nucleotides long and should be able to hybridize to the target DNA during the annealing phase of the PCR (*Figure 9.4*). The probe length may vary but should allow the dissociation from the target DNA at temperatures of 7–10°C higher than the annealing temperature. It is important to consider only the probe sequence of the molecular beacon and not the arms when calculating the T_m, because the arms are not involved in the hybridization event.

Once the probe sequence has been selected, two complementary arm sequences are designed and added on either side of the probe sequence. The two arm sequences allow molecular beacons to adopt a hairpin structure that forms a stable stem (*Figure 9.4*). The length of the arms may vary but short arms, containing 5–8 base pairs, have been shown to be stable in the presence of 1 mM $MgCl_2$ (3). Although the more economical solution is to design short arms, it is important that the arms have the correct length and DNA sequence to allow for a T_m similar to the probe (i.e. 7–10°C higher than the annealing temperature). It is advisable to design the arms with a 75–100% GC content, which will keep the length to a minimum; however, as for TaqMan probes a guanidine residue should not be present next to the fluorescent reporter as this quenches the fluorophore. Because the hairpin structure of a molecular beacon is formed by an intramolecular hybridization event the GC rule (Chapter 3) cannot be applied to calculate the T_m.

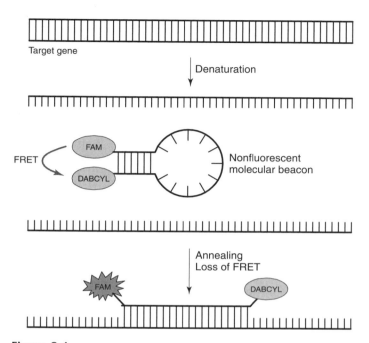

Target gene

Denaturation

FRET

FAM

DABCYL

Nonfluorescent
molecular beacon

Annealing
Loss of FRET

FAM

DABCYL

Figure 9.4

A schematic diagram showing the principle of molecular beacons as a reporter
during real-time RT-PCR. In addition to general PCR components the reaction also
contains a target gene-specific molecular beacon containing a fluorescent dye
(FAM) at the 5′-end and a quencher dye (DABCYL) at the 3′-end. Because the
molecular beacon adopts a hairpin stem structure when free in solution the
fluorescent dye and the quencher dye are in close proximity, allowing FRET to
occur resulting in no fluorescence. During the annealing step the hairpin structure
dissolves and the molecular beacon anneals to the target amplicon resulting in
loss of FRET and increased fluorescence ( ).

To calculate the T_m DNA folding software should be used, such as the Zuker
DNA folding program (http://www.bioinfo.rpi.edu/applications/mfold/ old/
dna/form1.cgi). As a rule of thumb a five base pair-long GC-rich stem will
melt between 55 and 60°C and a six base pair-long GC-rich stem will melt
between 60 and 65°C, whilst a seven base pair-long GC-rich stem will melt
between 65 and 70°C. The Zuker DNA folding program can also be used to
assess the probability of the free molecular beacon forming a hairpin structure rather than alternative structures. If alternative structures form, the
fluorescent reporter and the fluorescent quencher may not be placed in the
immediate vicinity, resulting in background fluorescence. Alternatively,
longer stems may form, resulting in slow binding to the target DNA.

In parallel with the molecular beacon design the amplification primers
should also be considered and it is important that there is no complementarity between the amplification primers and the molecular beacon.
This may cause the molecular beacon to hybridize to one of the primers,
resulting in primer extension by the DNA polymerase. It is also advisable
to design amplification primers that will result in an amplicon of approxi-

mately 150 base pairs as this will make the amplification reaction more efficient. In addition a shorter amplicon allows the molecular beacon to compete for its target more efficiently, which in turn will produce a stronger fluorescent signal. A useful review of molecular beacons is provided at http://www.bio.davidson.edu/courses/Molbio/MolStudents/spring2000/palma/beacons.html. Reagents are available from Sigma-Aldrich.

Although molecular beacons have the advantage of lower background fluorescence and greater specificity compared with TaqMan probes, they can be difficult to design and optimize. In light of this, Premier Biosoft International (http://www.premierbiosoft.com) have developed a software package that will design molecular beacons automatically. A free trial version can be downloaded from the Premier Biosoft International website; however, once the trial version expires the price of this software is in the region of 2 000 US dollars.

Scorpions

Scorpion probes are similar to molecular beacons in that they form a hairpin structure when free in solution; however, their design and mode of action differs substantially. Scorpion probes are bi-functional molecules that contain a PCR amplification primer covalently attached to a probe sequence. As for molecular beacons, the probe sequence is held in a hairpin configuration by complementary arm sequences (*Figure 9.5*), ensuring that the fluorescent reporter, such as FAM, and the fluorescent quencher, such as DABCYL, are in close proximity, allowing FRET to occur (*Figure 9.5*). However, in contrast to molecular beacons the hairpin structure of scorpions is attached to the 5′-end of a target gene-specific oligonucleotide primer (*Figure 9.5*). During the annealing and extension stage of the PCR the scorpion primer anneals to the target DNA and the primer is extended by *Taq* DNA polymerase to form an amplicon (*Figure 9.5*). To ensure that the DNA polymerase does not read through the scorpion primer, and by doing so copy the probe region, a nonamplifiable monomer (blocker) is added between the fluorescent quencher and the primer (*Figure 9.5*). After the extension phase and a second round of denaturation, the hairpin structure opens up, allowing the probe, which contains an amplicon-specific sequence, to curl back and hybridize to the target sequence in the PCR product (*Figure 9.5*). Because the hairpin structure requires less energy to denature than the newly formed DNA duplex, the sequence-specific probe hybridizes to the target amplicon with great speed and accuracy. The opening up of the hairpin loop prevents the fluorescence reporter from being quenched and an increase in fluorescence is observed.

The design of the amplicon-specific probe and the stem sequences in scorpions is essentially the same as for molecular beacons (see above). First, the probe sequence should have a T_m approximately 7–10°C lower than the T_m of the hairpin stem. Second, the stem sequences should be between five and eight base pairs, GC rich but avoiding a guanosine residue next to the fluorescent reporter. Third, the probe sequence should be between 15 and 30 nucleotides long. In addition to these considerations the target for the probe should not be more than 11 nucleotides from the 3′-end of the scorpion.

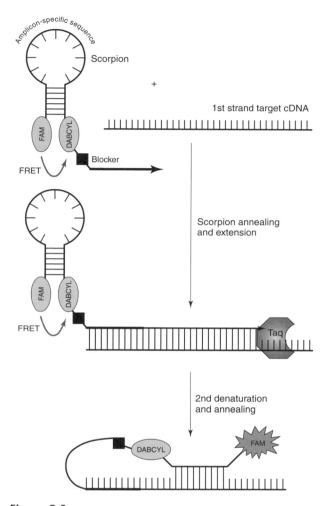

Figure 9.5

A schematic diagram showing the principle of a scorpion as a reporter during real-time RT-PCR. In addition to general components the reaction also contains a target gene-specific scorpion probe containing a fluorescent dye (FAM) at the 5'-end and a quencher dye (DABCYL) at the 3'-end. As for molecular beacons the scorpion adopts a hairpin stem structure when free in solution, ensuring that the fluorescent dye and the quencher dye are in close proximity. This allows FRET to occur, resulting in no fluorescence. In contrast to molecular beacons the hairpin structure of a scorpion probe is attached to the 5'-end of a target gene-specific oligonucleotide primer. During the PCR annealing and extension stage the scorpion primer anneals to the target DNA and the primer is extended by *Taq* DNA polymerase to form an amplicon. To ensure that the DNA polymerase does not read through the scorpion primer a nonamplifiable blocker (■) is placed between the fluorescent quencher and the gene-specific primer. After a second round of denaturation, the hairpin structure dissolves. This allows the probe, which contains an amplicon-specific sequence, to curl back and hybridize to the target sequence in the PCR product, resulting in loss of FRET and increased fluorescence (FAM).

There are several advantages of scorpions over molecular beacons. In contrast to molecular beacons and the TaqMan approach, scorpion assays only require one probe. Because the tail of the scorpion and the PCR product is part of the same DNA strand the reaction leading to a fluorescent signal is unimolecular, in sharp contrast to the bimolecular collisions needed by TaqMan and molecular beacons. The scorpion unimolecular arrangement effectively means that the reaction leading to signal generation occurs almost spontaneously, before any competing reactions such as target amplicon reannealing. This in turn leads to stronger signal generation and less background fluorescence. Furthermore, because of the rapid reaction kinetics there is no need for long extension periods. The unimolecular arrangement also results in increased discrimination and specificity and scorpions are ideal for SNP testing (Chapter 11). Although scorpions are relatively expensive to make they do not suffer from temperamental probe design as do molecular beacons. Scorpions are also compatible with any commercial fluorescent dye and, together with the low background fluorescence, extended multiplexing can be performed combining four or more separate reactions.

LightUp probes

LightUp probes (LightUp Technologies, Sweden) represent the simplest method for monitoring amplicon accumulation during real-time RT-PCR. LightUp probes are based on peptide nucleic acid (PNA), which is a mimic of DNA (4) that contains an asymmetric cyanine dye (5). Upon PNA probe hybridization the dye binds to the target amplicon, resulting in a dramatic increase in fluorescence. The increase in fluorescence is so great that it can be seen by the naked eye in a test tube.

PNA consists of a N-(2-aminoethyl)glycine backbone to which the bases are attached and PNA hybridizes to DNA and RNA in an anti-parallel fashion following standard Watson and Crick rules. There are several advantages of PNA over DNA. First, DNA hybridization requires the presence of salt to balance the phosphate backbone negative charge whilst PNA does not require salt due to its neutral backbone. Second, PNA has greater sequence specificity than DNA, which allows the use of shorter PNA oligomers as probes. Third, single-stranded PNA is completely resistant to enzymes such as nucleases, peptidases and proteases, which commonly degrade DNA. LightUp probes take advantage of these facts and although their use in real-time RT-PCR is not yet extensive (probably due to their relatively high cost), they have several benefits over other amplicon detection systems. Since LightUp probes contain only a single fluorescent dye a simple increase in fluorescence is measured, in contrast to a change in fluorescence energy distribution when using energy transfer probes such as TaqMan probes, molecular beacons and scorpions. In addition, because LightUp probes fluoresce after hybridization to the target amplicon, they can be designed to bind only during the annealing step and not during extension. This in turn does not interfere with the PCR as is the case when using TaqMan probes which are degraded by the *Taq* DNA polymerase (see above). The use of a single dye that fluoresces immediately after hybridization also eliminates the need for conformational changes to take place for signal generation, as is the case for molecular beacons and scorpions. Taking all these considerations together, LightUp probes are easy to design and require little optimization and are likely to become more commonly used.

9.4 General guidelines for probe and primer design

Although primer dimers and nonspecific amplification products will not be detected when using fluorescent probe approaches, they will affect the efficiency and dynamics of the PCR. Because of this there are a number of general guidelines that should be followed when designing fluorescent probes and oligonucleotide primers for real-time RT-PCR experiments. The following paragraphs outline general rules regarding probe and primer design, amplicon characteristics and fluorophore selection.

Probe design

The consensus amongst researchers setting up real-time RT-PCR experiments is that it is generally easier to design the fluorescent probe sequence first and then the amplification primers.

The length of a fluorescent probe should be between 18 and 30 nucleotides, with an optimal length of 20 nucleotides. Although it is possible to use probes with more than 30 nucleotides it is recommended that in such cases the fluorescent quencher is not placed at the 3'-end but internally between 18 and 25 nucleotides from the 5'-end. This is only an issue when using TaqMan probes and this ensures efficient energy transfer from the fluorescent reporter to the fluorescent quencher minimizing background fluorescence. Probes should also have a GC content between 30 and 80% however, it is more favorable to have more cytosine than guanosine so the correct strand should be selected accordingly. Related to this, probes should not contain runs of identical nucleotides, and runs of four or more guanidine residues should be avoided at all costs. A guanosine at the very 5'-end of the probe should also be avoided as this will quench the fluorescent dye.

The T_m of the probe should be between 8 and 10°C higher than the T_m of the amplification primers. As a general rule of thumb an 8°C higher T_m should be used for genotyping whilst a 10°C increase should be used for gene expression profiling. It is also important when designing a probe to avoid any mismatches between the probe and the target, as well as avoiding any complementarity with either of the amplification primers.

In terms of probe placement it is advisable to design the probe so that its 5'-end hybridizes as close to the 3'-end of the forward amplification primers as possible without actual overlapping. This again is only crucial for TaqMan probes as this ensures an optimal 5'-nuclease reaction during probe displacement.

The general design rules described above should be used for both singleplex and multiplex reactions; however, when designing probes for multiplex reactions the position of the polymorphism should be in the middle of the probe and the T_m should be the same for both probes.

Amplification primers

Once the probe sequence has been designed the forward and reverse amplification primers should be designed. The amplification primers should be designed following the standard rules as described in Chapter 3, although primer design for real-time RT-PCR approaches is perhaps more important than for standard analytical PCR. The T_m of the forward and reverse primer should be between 63 and 67°C, depending on the T_m of the probe (see above). It is important to avoid runs of identical nucleotides and to avoid complementarity within the primers, as this would result in hairpin formation and less efficient amplification. It is also important to avoid complementarity between the primers so as to avoid primer-dimer formation. One problem often encountered during real-time RT-PCR is the amplification and detection of contaminating genomic DNA. This can be avoided by designing primers that span or flank introns. For intron-spanning primers the first half of the primer should hybridize to the 3′-end of one exon and the 5′-end of the other exon, which will only allow cDNA amplification to take place. For intron-flanking primers the forward and reverse primers should hybridize to separate exons. By using such primers amplicons from cDNA will be smaller than amplicons from genomic DNA and the difference in amplicon size can be determined from melting curve analysis.

Amplicon characteristics

When designing the forward and reverse primers it is important to take into consideration amplicon size. The length of the amplicon should be between 80 and 120 base pairs since the shorter the amplicon the more efficient the amplification reaction. In addition, a small amplicon size also increases the efficiency of the nuclease reaction associated with the use of TaqMan probes. Another important factor is the GC content, which should ideally be between 40 and 60%, although amplicons with as low as 30% or as high as 80% GC content will also work. During the amplification reaction it is crucial to avoid secondary structure formation and the amplicon sequence should therefore be analyzed in terms of possible secondary structure elements.

9.5   Instruments and quantification of results

There are a range of real-time PCR instruments and the properties of some of these are summarized in *Table 9.2*.

The quantification of results from a real-time RT-PCR experiment is important and there are two ways commonly used to do this: the standard curve method and the comparative threshold or C_T method. Both methods have different strengths and will be discussed below.

Standard curve method

The standard curve method is based on using a sample of RNA or DNA of known concentration to construct a standard curve. Once a standard curve has been generated it can then be used as a reference standard for the extrapolation of quantitative information regarding mRNA targets of

Table 9.2   Real-time thermal cyclers

PCR system	Company	Excitation specifications	Detection specification	Sample format
Applied Biosystems 7300 Real-Time PCR System	Applied Biosystems	Tungsten Halogen fixed wavelength	FAM™/SYBR® Green 1, VIC™/JOE, NED™/ TAMRA™, ROX™ dyes	96-well plates 0.2 ml tubes
Applied Biosystems 7500 Real-Time PCR System	Applied Biosystems	Tungsten Halogen variable wavelength	FAM™/SYBR Green 1, VIC™ /JOE, NED™/ TAMRA™/ CY3 Dye™, ROX™/Texas Red , CY5 Dye™	96-well plates 0.2 ml tubes
Exicycler™ Real-Time Thermal Block	Bioneer	Metal halide lamp, filters: 4 channels (480–640 nm)	4 channels (520–680 nm)	96-well plates
iCycler iQ Real-Time PCR Detection System	Bio-Rad	400–700 nm	CCD with intensifier technology	96-well plates
MyiQ Single-Color Real-Time PCR Detection System	Bio-Rad	400 nm FAM/SYBR	585 nm FAM/SYBR	96-well plates
Smart Cycler® II System	Cepheid	4 channels (450–495, 500–550, 565–590, 630–650 nm)	4 channels (510–527, 565–590, 606–650, 670–750 nm)	16 programmable reaction sites
R.A.P.I.D. ™ System	Idaho Technology Inc.	450–490 nm	3 channels (520–540, 630–650, 690–730 nm)	32-well plates
Chromo 4™ Four-Color Real-Time System	MJ Research	4 channels (450–490, 500–535, 555–585, 620–650 nm)	4 channels (515–530, 560–580, 610–650, 675–730 nm)	96-well plates 0.2 ml tubes 8-tube strips
DNA Engine Opticon® Continuous Fluorescence Detection System	MJ Research	450–495 nm	515–545 nm	96-well plates 8-tube strips
DNA Engine Opticon® 2 Continuous Fluorescence Detection System	MJ Research	470–505 nm	2 channels (523–543, 540–700 nm)	96-well plates 8-tube strips
LightCycler® Instrument	Roche Applied Science	Blue LED light 470 nm	3 channels (530, 640, 710 nm)	32-well plates
LightCycler® 2.0 Instrument	Roche Applied Science	Blue LED light 470 nm	6 Channels (530, 560, 610, 640, 670, 710 nm)	32-well plates 20 µl reaction cuvettes

Table 9.2 *continued*

PCR system	Company	Excitation specifications	Detection specification	Sample format
Mx3000P™ Real-Time PCR System	Stratagene	Quartz Tungsten Halogen lamp 350–750 nm	1 scanning photomultiplier tube (PMT)	96-well plates
Mx4000™ Multiplex Quantitative PCR System	Stratagene	350–750 nm	350–830 nm	96-well plates

unknown concentration. There are a variety of different standards that can be used, including *in vitro* reverse transcribed mRNA, *in vitro* synthesized single-stranded DNA, or purified plasmid DNA. To generate the standard curve the concentration of the DNA or RNA standard should be measured using a spectrophotometer (260 nm) followed by conversion into the number of copies using the molecular weight of the sample. Although DNA is most commonly used as a standard, absolute quantification of mRNA expression requires the use of *in vitro* reverse-transcribed RNA because this accounts for the efficiency of the reverse transcription step. However, using reverse-transcribed RNA to construct a standard curve is labor intensive because of the need to generate a cDNA plasmid for the *in vitro* reverse transcription reaction. Another potential problem when using *in vitro* reverse-transcribed RNA relates to RNA stability and this should be taken into account when determining the concentration.

By far the most commonly used standards are cDNA plasmids. Although cloning of cDNAs into plasmid vectors can be time-consuming, the advantage is that this now represents an unlimited source of a DNA standard. Because large amounts of plasmid DNA can be generated with ease, numerous experiments can be performed using the same dilutions of the same standard, which will minimize variations between different assays. Despite this advantage, cDNA plasmid standards will not account for variations during reverse transcription and will therefore only provide information on relative changes in mRNA expression. This is however not a large problem in that this and variations in RNA input amounts can be corrected for by normalization to housekeeping genes (Section 9.6).

The comparative threshold (C$_T$) method

The comparative threshold (C$_T$) method is based on the relative quantification of transcript levels and involves the comparison of C$_T$ values of the samples being analyzed with those of a control (calibrator) such as for example nontreated sample or RNA from an untreated sample. As explained in *Table 9.1* the threshold cycle (C$_T$) indicates the cycle number where the reaction fluorescence crosses the threshold, which reflects the reaction point when sufficient amplicons have been generated to give a significant fluorescent signal over the baseline. The equation used to calculate relative gene expression levels using the comparative C$_T$ method is given by $2^{-\Delta\Delta CT}$ where $\Delta\Delta C_T = ((\Delta C_T \text{ of the target sample}) - (\Delta C_T \text{ of the calibrator}))$. The ΔC_T

is the C_T of the target gene (or the calibrator) subtracted from the C_T of a housekeeping gene. The described equation therefore represents the normalized (Section 9.6) expression of the target gene relative to the normalized expression of the calibrator. For the $\Delta\Delta C_T$ equation to be valid the PCR efficiency of the target gene has to be more or less equal to the efficiency of the housekeeping gene and this is often not the case. If the efficiency of the target gene is not the same as for the housekeeping gene the comparative C_T method cannot be used and the standard curve method should therefore be used (see above). The need for equal efficiencies for both the target and housekeeping genes is clearly a disadvantage of this method; however, in contrast to the standard curve method there is no need to construct standards and moreover the entire 96-well microtiter plate can be used for target samples.

9.6 Normalization and control selection

In order to obtain reliable quantitative results from a real-time RT-PCR experiment, variations in both the reverse transcription reaction and the PCR itself must be corrected for. Differences in the efficiency of the reverse transcription reaction will result in a cDNA amount that does not correspond to the amount of RNA and a difference in PCR amplification efficiency will result in differences in amplicon accumulation. The latter is not a major problem in that quantification is based on C_T values determined early in the exponential phase of the PCR reaction.

The most acceptable way of correcting for differences in input RNA and reverse transcription efficiencies is based on normalization of the target gene to a housekeeping gene. A number of housekeeping genes can be used but it should be expressed at a constant level in all tissues and at all stages of development. In addition, the treatment of the unknown sample should not affect the housekeeping gene. The commonly used housekeeping genes are glyceraldehyde-3-phosphate dehydrogenase (GAPDH) or β-actin.

GAPDH is a very abundant glycolytic enzyme involved in a number of cellular processes which is present in most tissues and has been used extensively for normalization purposes. However, a number of studies have shown that GAPDH is in fact not suitable as an internal control because its expression is affected by a number of treatments (6,7).

β-actin is a cytoskeletal protein and is expressed in almost all tissue types. Although its expression can vary under some conditions and in some diseases β-actin probably represents the best housekeeping gene for use in real-time RT-PCR experiments. Studies have shown that β-actin levels do not vary under numerous treatments in a variety of samples (8).

Although β-actin fulfils the criteria of a good housekeeping gene the optimal choice of a housekeeping gene is dependent on the experiment in question and should therefore be selected with this in mind.

9.7 A typical real-time RT-PCR experiment using SYBR® Green I

By reading the previous Sections in this Chapter you should now be familiar with the theory behind real-time RT-PCR. However, to give an idea of the more practical aspects of the technique and to become more familiar with data sets a typical real-time RT-PCR experiment together with a typical data set using SYBR® Green I is outlined below. The experiment made use of an MJ Research PTC 200 thermocycler with a Chromo4 Continuous Fluorescence Detection system and data analysis was performed using MJ Research Opticon Monitor v. 2.03.5 software. The experiment described involves analysis of the *NADPH: protochlorophyllide oxidoreductase* (*PORA*) gene in the model plant *Arabidopsis thaliana* in response to far-red light irradiation. In wild-type *Arabidopsis* seedlings the expression of *PORA* rapidly decreases when seedlings are transferred from total darkness to far-red light conditions. By contrast, the *fhy1 Arabidopsis* mutant, affected in a number of far-red responses, shows defective *PORA* gene expression in response to far-red light irradiation. Real-time RT-PCR was used to quantify the difference in *PORA* gene expression between wild-type (Laer-Landsberg *erecta*) and the *fhy1* mutant after exposure to 6 and 18 hours of far-red light.

Wild-type and *fhy1 Arabidopsis* seedlings were grown for 5 days in the dark followed by transfer to monochromatic far-red light treatment. Tissue was harvested after 0, 6, and 18 hours of far-red light treatment and cDNA was synthesized using an oligo-dT primer. For the PCR amplification oligonucleotide primers were designed to the 3′-end of the *PORA* gene and oligonucleotide primers for β-actin were used as an internal reference/ control (Section 9.6).

When using SYBR® Green I as a detection method the first step is to test the efficiency of the oligonucleotide primers (Section 9.3). This not only gives a measure of amplification efficiency but also enables one to check that only one amplification species is being produced. The efficiency is calculated using a cDNA dilution series and calculated by $E = 10^{-1/\text{slope}}$ (9). As observed from *Figure 9.6(A)* the primers designed for the *PORA* amplification give rise to a single defined fluorescence peak demonstrating that these primers are well suited for real-time quantification.

Once it has been demonstrated that the oligonucleotide primers are specific for the gene in question the real-time experiment can be performed. In this case real-time PCR was used to amplify β-actin and *PORA* from the different individual experimental samples. Three replicates of each experimental sample were run together with a water control and a minus reverse transcription (RT) control (to eliminate DNA contamination). A threshold value (Section 9.2 and *Table 9.1*) was chosen above the fluorescence background (*Figures 9.6(B)* and *9.1*) and any fluorescence signal over this threshold was measured, which determines the C_T values. The C_T values were recorded for each light treatment (far-red and darkness) from both wild-type and *fhy1* samples. In order to calculate the relative expression of *PORA* in far-red light as compared with darkness the C_T values were used in the following equation:

$$R = \frac{(E_{\text{target}})^{\Delta\text{CPtarget (control–sample)}}}{(E_{\text{reference}})^{\Delta\text{CPreference (control–sample)}}}$$

The relative value (R) of each time point in relation to wild-type grown in darkness was then plotted against hours exposed in far-red light. As can be seen from *Figure 9.6(C)* the real-time RT-PCR data show that *PORA* expres-

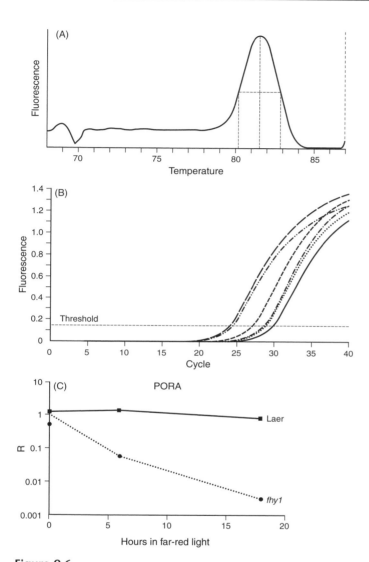

Figure 9.6

An example of a real-time RT-PCR experiment using SYBR® Green I as reporter. The aim of the experiment was to determine the relative difference in the expression level of the *NADPH: protochlorophyllide oxidoreductase* (*PORA*) gene in wild-type and *fhy1* mutant *Arabidopsis* seedlings in response to 0, 8, and 18 h of far-red light irradiation. (A) Melting curve analysis using the *PORA*-specific primers showing a single, well-defined peak demonstrating that only one amplification product is being produced. (B) During the first 3–15 amplification cycles a threshold baseline is determined and the fluorescence above this threshold is measured giving the C_T values for each reaction/sample. (C) From the C_T values the relative expression (R) of *PORA* is calculated for both wild-type and the *fhy1* mutant at each time-point and in this case plotted against hours exposed to far-red irradiation. This demonstrates that *PORA* expression decreases in wild-type but remains constant in *fhy1* seedlings. Data kindly provided by Dr Trudie Allen, (University of Leicester).

sion in wild-type seedlings decreases upon exposure to far-red light whilst in the *fhy1* mutant *PORA* expression remains unchanged.

This simple example demonstrates the power of real-time RT-PCR in determining and comparing relative expression levels of a gene in wild-type and in a mutant under two different environmental conditions.

9.8   Common real-time RT-PCR pitfalls

When designing and performing any PCR amplification there are a number of potential pitfalls and real-time RT-PCR is no exception. Most pitfalls are easy to solve and the most common pitfalls and problems are described below.

Primer and probe design

If the oligonucleotide primers and/or the probe are poorly designed your real-time RT-PCR experiment will have limited success if any. Although amplification primers for everyday PCR amplification can be designed manually using general design rules (Chapter 3), it is recommended that primer design software is used to design both the amplification primers and the probe for your real-time RT-PCR experiment (Section 3.7). Most design programs have a number of adjustable parameters including primer/probe T_m, complementarity, and secondary structure.

RNA quality

The quality of your RNA is extremely important in order to ensure successful real-time RT-PCR. Degraded or impure RNA will limit the efficiency of the RT reaction, reducing yield, and thereby not giving an accurate representation of gene expression.

Minus RT control

To completely eliminate genomic DNA from RNA preparations is virtually impossible. Because of this a mock reverse transcribed RNA sample, containing all reagents apart from the RT, should be included in the real-time experiment. No fluorescence increase would demonstrate no DNA contamination. To overcome problems with DNA amplification, probes and primers can be designed to span exon–intron borders, however this is not always possible.

Normalization control

To correct for sample-to-sample variation during real-time RT-PCR an invariant endogenous control gene should be included. The most common controls include β-actin and GAPDH (Section 9.6).

Melting curve analysis

A common mistake is to avoid performing melting curve analysis when using SYBR® Green I as a reporter. Ideally a single and well-defined peak should be observed at the melting temperature of the amplicon which demonstrates that the fluorescence is a direct measure of specific product accumulation. If a series of peaks are observed the discrimination between specific and nonspecific product formation is too small, which requires optimization of the PCR itself.

Fluorescence baseline and thresholds

To obtain accurate C_T values the baseline has to be set at least two cycles before the first C_T value of the most abundant sample. The threshold should be in relation to the baseline and set when the products reach exponential phase approximately 10 deviations above the baseline.

Reaction efficiency

The reaction efficiency (E) is calculated by the equation $E = 10^{-1/\text{slope}}$ (9) and the efficiency of the PCR should be between 90% and 110%. If the reaction efficiency is lower than 90% optimization is needed and you should start considering the amplicon size, secondary structures and primer design.

9.9   Applications of real-time RT-PCR

Since the advent of real-time RT-PCR the number of applications has grown exponentially. It is outside the scope of this book to describe every application of real-time RT-PCR; however we will describe two general but important application areas below.

Cancer research and detection

The number of patients with leukemia or lymphoma who achieve a complete remission after initial treatment is very high. However, many patients will relapse due to residual and undetected tumor cells. For the detection of rare tumor cells in clinical samples, real-time RT-PCR offers two main advantages over conventional RT-PCR. First, the results are quantitative, which gives a measure of the number of residual tumor cells, and second, it facilitates exact sensitivity controls on a per-sample basis. In addition real-time RT-PCR is used more frequently in detecting the molecular events underlying disease reoccurrence, which may guide therapeutic decisions.

When using peptide-based anticancer vaccines it is important to analyze circulating lymphocytes to enumerate the number of T cells elicited by the treatment. However, it is also important to gain information regarding their functional state. Real-time RT-PCR has been used to assess directly the immune status of peripheral blood mononuclear cells (PBMC) from melanoma patients being treated with peptide-based vaccines (10). By testing PMBC for *interferon-γ* mRNA expression the PMBC response to vaccines could be assessed in a direct way. Another exciting application of real-time RT-PCR involves analysis of dynamic changes in expression of tumor-associated antigens and cytokine expression during immunization (11).

Real-time RT-PCR can also be used to screen for genomic mutations and polymorphisms. For example, the identification of *BRCA1* and *BRCA2* mutations permits molecular diagnosis for breast cancer susceptibility, and different types of fluorescent molecular probes have been used to successfully perform allelic discrimination (11).

Real-time RT-PCR and plants

The detection and quantification of foreign DNA present in plants is becoming increasingly important. Plants are subject to a myriad of infections, and real-time RT-PCR has been used to measure pathogen infection (12) in the field and for applied purposes. For example, seed potatoes cannot be sold within the European Union (EU) if they contain the potato brown rot agent *Ralstonia solanacearum*. In contrast to lengthy pathogenicity tests on tomato seedlings, real-time RT-PCR can quantitatively detect *R. solanacearum* rapidly (13).

The contamination of food with mycotoxins is of great concern since many are carcinogenic. However, toxin abundance does not correlate with fungal contamination but with the toxinogenic properties of individual strains. Real-time RT-PCR has been used to detect genes involved in toxinogenesis in the tricotecene-producing *Fusarium* (14) and in the aflatoxin-producing *Aspergillus* species (15).

With the introduction of stringent food safety regulations, real-time RT-PCR has seen increased use in assessing foreign DNA contamination in processed food. For instance, the absence of gluten in baby food is controlled by real-time RT-PCR amplification of cereal genes (16). Similarly, real-time RT-PCR is used to assess common wheat adulteration in durum wheat pasta (17) since Italian, Spanish and French regulations only allow up to 3% common wheat contamination in pasta and semolina.

References

1. Heid CA, Stevens J, Livak KJ, Williams PM (1996) Real time quantitative PCR. *Genome Res* **6**: 986–994.
2. Gibson UE, Heid CA, Williams PM (1996) A novel method for real time quantitative RT-PCR.. *Genome Res* **6**: 995–1001.
3. Tyagi S, Kramer FR (1996) Molecular beacons: probes that fluoresce upon hybridization. *Nature Biotechnology* **3**: 303–308.
4. Egholm M, Buchardt O, Christensen L, Behrens C, Freier SM, Driver DA, Berg RH, Kim SK, Norden B, Nielsen PE (1993) PNA hybridizes to complementary oligonucleotides obeying the Watson–Crick hydrogen-bonding rules. *Nature* **365**: 566–568.
5. Svanvik N, Westman G, Wang D, Kubista M (2000) Light-up probes: thiazole orange-conjugated peptide nucleic acid for detection of target nucleic acid in homogeneous solution. *Anal Biochem* **281**: 26–35.
6. Suzuki T, Higgins PJ, Crawford DR (2000) Control selection for RNA quantitation. *BioTechniques* **29**: 332–337.
7. Ke LD, Chen Z, Yung WK (2000) A reliability test of standard-based quantitative PCR: exogenous vs endogenous standards. *Mol Cell Probes* **14**: 127–135.
8. Giulietti A, Overbergh L, Valckx D, Decallonne B, Bouillon R, Mathieu C (2001)

An overview of real-time quantitative PCR: applications to quantify cytokine gene expression. *Methods* **25**: 386–401.

9. Pfaffl M (2001) A new mathematical model for relative quantification in real-time RT-PCR. *Nucleic Acids Res* **29**: e45.

10. Panelli MC, Wang E, Monsurro V, Marincola FM (2002) The role of quantitative PCR for the immune monitoring of cancer patients. *Expert Opin Biol Ther* **2**: 557–564.

11. Mocellin S, Rossi CR, Pilati P, Nitti D, Marincola FM (2003) Quantitative real-time PCR: a powerful ally in cancer research. *Trends Mol Med* **9**: 189–195.

12. Gachon C, Mingam A, Charrier B (2004) Real-time PCR: what relevance to plant studies? *J Exp Bot* **55**: 1445–1454.

13. Weller SA, Elphinstone JG, Smith NC, Boonham N, Stead DE (2000) Detection of *Ralstonia solanacearum* strains with a quantitative, multiplex, real-time, fluorogenic PCR (TaqMan) assay. *Appl Environ Microbiol* **66**: 2853–2858.

14. Schnerr H, Niessen L, Vogel RF (2001) Real time detection of the *tri5* gene in *Fusarium* species by lightcycler-PCR using SYBR Green I for continuous fluorescence monitoring. *Int J Food Microbiol* **4**: 53–61.

15. Mayer Z, Bagnara A, Farber P, Geisen R (2003) Quantification of the copy number of *nor-1*, a gene of the aflatoxin biosynthetic pathway by real-time PCR, and its correlation to the cfu of *Aspergillus flavus* in foods. *Int J Food Microbiol* **82**: 143–151.

16. Sandberg M, Lundberg L, Ferm M, Yman IM (2003) Real-time PCR for the detection and discrimination of cereal contamination in gluten free foods. *European Food Research and Technology* **217**: 344–349.

17. Alary R, Serin A, Duvaiu MP, Joudrier P, Gautier MF (2002) Quantification of common wheat adulteration of durum wheat pasta using real-time quantitative polymerase chain reaction (PCR). *Cereal Chem* **79**: 553–558.

Cloning genes by PCR

10

The cloning of genes is often a crucial step in a scientific project and can be both difficult and time-consuming. The use of PCR has greatly enhanced the successes of gene isolation. Cloning of genes by PCR can be divided into two main areas: (i) genes of known DNA sequence; and (ii) genes of unknown DNA sequence. Genome sequencing projects (Chapter 11) are generating an increasing amount of data that makes cloning of genes more straightforward, however there remain many cases where unknown genes must be cloned. This Chapter deals with the cloning of both unknown genes and those that have been previously isolated.

A   Cloning genes of known DNA sequence

10.1   Using PCR to clone expressed genes

If a DNA fragment has been isolated containing part of the target gene, perhaps as a genomic sequence, it can be used to clone a full-length cDNA for further analysis. Perhaps quantitative RT-PCR (Chapter 8) or real-time RT-PCR (Chapter 9) has indicated very low levels of expression of the target gene, and hybridization screening of a cDNA library, using the isolated fragment as a probe, fails to yield clones. Dealing with a low-abundance transcript can often be frustrating as conventional cDNA library screening is labor intensive and success depends on a number of parameters associated with the quality of the library. First, the quality of the mRNA used to generate the cDNA library is of great importance since low-abundance transcripts can easily be 'lost' during sample handling. Second, the efficiency of the first- and second-strand cDNA synthesis should be optimized and monitored by incorporating radiolabeled nucleotides. Third, the proportion of recombinant clones should be as high as possible to reduce the number of plaques or bacterial colonies needed to be screened and to increase the likelihood of cloning low-abundance transcripts. Even if you manage to generate a good cDNA library it may not be possible to isolate certain low-abundance cDNA clones. By contrast, it is often possible to isolate such cDNAs using PCR-based techniques.

Generating cDNA libraries by PCR

Various approaches have been applied to the construction of cDNA libraries by PCR. Often the rationale for using such an approach is the limited amount of material available from which mRNA can be produced. Due to the limitations on materials, such procedures rely on the use of total RNA preparations as the source of templates for mRNA reverse transcription and cDNA amplification. An inevitable consequence of this strategy is the amplification of rRNA sequences that predominate in any total RNA preparation and which form templates for nonspecific or self-priming reactions leading to a reduction in library quality.

Early methods were based on an oligo-dT primer for first-strand cDNA synthesis and homopolymer-tailing, often by dCTP, of the 3′-end of these cDNA strands. PCR with oligo-dG and oligo-dT primers was then performed. This approach was improved by the inclusion of specific sequence extensions on the oligo-dG and oligo-dT primers so that rather than using the homopolymer tracts as priming sites, specific primers complementary to the primer extensions could be used for increased specificity. Alternatively, and more efficient than homopolymer tailing, following standard double-strand cDNA synthesis the molecules can be blunt-ended by treatment with, for example, Klenow fragment and dNTPs, and a double-stranded adaptor ligated to provide specific priming sites. Of course in this case the new priming site would be added to both the 5′- and 3′-ends of the cDNA allowing amplification by a single primer, but this also results in single strands that have complementary ends that are capable of annealing. The consequence is a process called suppression, which results in such self-associated molecules being unavailable as templates for PCR. This suppression phenomenon has been exploited in some cDNA synthesis protocols to prevent the nonspecific amplification of rRNA sequences that are commonly recovered during cDNA library construction from total RNA preparations (1). In essence the procedure is identical to the generation of a library by ligation of a double-strand adaptor. The adaptor is added to the 5′- and 3′-ends of each molecule in the library whether derived from mRNA or rRNA. In the PCR step, however, the adaptor-base primer is added together with an oligo-dT primer. This will allow amplification of any molecule, but only the mRNA molecules that have a polyA tail will provide sites for both the adaptor and oligo-dT primer. Any molecules that are amplified only by the adaptor primer will have complementary terminal sequences that will be able to anneal, thus preventing the primer accessing the site and therefore suppressing the level of representation of such molecules in the final library. This provides an efficient method for the selective amplification of mRNA-derived cDNAs.

Solid-phase procedures for library construction have also been developed that either depend upon the capture of mRNA molecules, by annealing of the polyA tail to oligo-dT coupled to some form of solid support, or the use of a biotinylated oligo-dT primer for first-strand cDNA synthesis.

PCR amplification from a cDNA library

A cDNA library is a highly complex mixture of nucleic acids and often, in the case of a phage library, protein components, and so it is important to use high stringency conditions for the PCR reaction in order to minimize nonspecific background amplification. It is convenient to use PCR as a tool to rapidly screen random clones to determine the quality of a cDNA library. Essentially random plaques are transferred with a toothpick to a PCR mix and universal primers flanking the cDNA cloning region are used to amplify the inserts. A good library should give a high number of clones with inserts of varying sizes. An example of PCR screening of random clones from a bacteriophage λ gt10 cDNA library is shown in *Figure 10.1*.

For the isolation of target genes there are two general approaches to PCR amplification from a cDNA library:

- from the starting cDNA, which may be one of the increasing sources of commercially available PCR-ready cDNA samples specifically produced for this purpose; or
- from the phage library suspension.

During cDNA library construction (ligation, packaging, transfection) to yield the primary library and its subsequent amplification, the distribution of clones can be skewed such that the library is not representative of the starting mRNA population. This can have a particularly adverse effect on the representation of clones representing low-abundance transcripts. For this reason it is better, where possible, to start from a cDNA template source, since this increases the chance of isolating rare transcripts due to the higher complexity of cDNAs whilst reducing nonspecific amplification due to the lack of phage DNA. There are no major difficulties associated with direct PCR amplification from cDNA although the following points should be considered. First, use a low template concentration such as for genomic PCR, in the range of 10–50 ng, and second, for rare transcripts use 40–45 amplification cycles. Alternately, use 30 cycles followed by re-amplification of an aliquot for an additional 25 cycles.

SMART cDNA cloning

Clontech's SMART™ PCR cDNA synthesis kit facilitates production of high-quality cDNA from total or polyA RNA as shown in *Figure 10.2*. Reverse transcriptase uses a modified oligo-dT primer to generate first-strand cDNA. Upon reaching the 5′-end of the mRNA the terminal transferase activity of the reverse transcriptase adds additional nucleotides, normally deoxy-

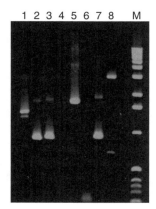

Figure 10.1

Screening random λgt10 plaques from a library for the presence of inserts. Several clones carry inserts of differing size (1, 2, 3, 5, 7, 8) while other clones show no apparent inserts (4, 6). Photography kindly provided by A. Neelam (University of Leeds).

cytidine, to the 3'-end of the first-strand cDNA. The SMART II oligonucleotide, containing a 3' oligo-G sequence, base pairs to these Cs on the
cDNA, and now acts as a 'new' template for the reverse transcriptase, which
extends the cDNA to the end of the SMART II oligonucleotide. The
extended full-length single-stranded cDNA, now containing two priming
sites (5' and 3'), can be used for end-to-end cDNA amplification by PCR.
The majority of cDNAs should represent full-length copies allowing for
efficient amplification of 5'-regions.

It is advisable for all cDNA library production schemes to use primer pairs
that contain engineered restriction sites that will facilitate subsequent
cloning of the PCR-amplified cDNA (Chapter 6).

The second option is to PCR from a phage cDNA library suspension. This
may result in more nonspecific amplification compared with direct PCR
from cDNA. When dealing with phage suspensions it is important to allow
access to the packaged DNA by heating an aliquot of the phage suspension
to 95°C for 5 min or by placing in a microwave oven for 5 min at full power
(700 W). As for direct PCR amplification from library DNA, a low concentration of template DNA should be used to minimize nonspecific

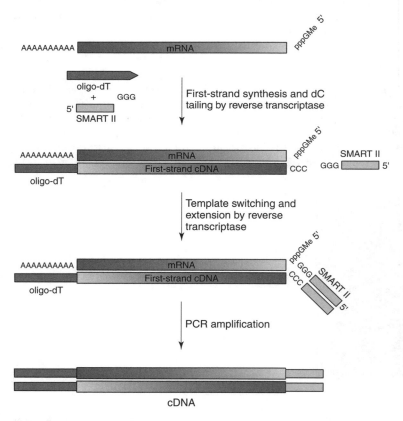

Figure 10.2

The principle of SMART™ PCR cDNA synthesis kit for generating full–length cDNA
molecules.

amplification events. When a cDNA library is generated it is usual to check the integrity of the library by analyzing random clones for the presence of inserts of varying sizes that correspond to different initial transcripts, and such a screen is shown in *Figure 10.1*. The identification of positive clones is usually achieved by filter transfer of plaques from a plate, followed by fixing the released DNA to the membrane, then hybridization with a labeled probe. In initial library screens it is difficult to isolate single plaques and so the screening must be repeated. However, PCR screening can be used to try to isolate individual clones by amplification from dilutions of a library (*Figure 10.3*). When the lowest dilution that still gives a positive result is identified this corresponds to the number of plaques that must be screened to isolate a single positive. If this number is small (10–50), then it is possible to pick individual plaques to screen. If the number remains large (>50) then a further hybridization experiment is probably more efficient.

10.2 Expressed sequence tags (EST) as cloning tools

DNA sequence databases provide a wealth of EST sequences and these can be used as very efficient tools for gene cloning by PCR. ESTs are DNA sequences of the 5′- or 3′-ends of cDNA clones often randomly picked from a cDNA library, or as a subpopulation of clones isolated from a developmental library, perhaps by differential screening. The sequence information is limited to usually about 500 nucleotides, the amount generated from a single sequencing reaction. Thus for any given cDNA clone there can be two ESTs, one corresponding to 5′- and one to 3′-sequence, but in many cases the region between these extremes is unknown. Nonetheless, the limited sequence information is sufficient to search databases to identify homology to known genes, or genomic regions. Most importantly, if you search a database with a sequence of interest and identify an EST, then this means that a cDNA clone of your target gene is available. In most cases

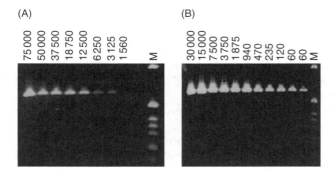

Figure 10.3

Screening dilutions of an enriched λgt10 cDNA library for the presence of a target clone. The number of plaque–forming units (p.f.u.) present in the PCR are indicated above each lane; M is molecular size markers. (A) The initial enrichment shows detection of a clone in 6 250 p.f.u. (B) Subsequent enrichment reveals the presence of a clone in the highest dilution sample that contains 30 p.f.u. Photographs kindly provided by A. Neelam (University of Leeds).

ESTs can be ordered, for a small handling fee, from various stock centers in the form of a plasmid containing the cDNA. There are also a growing number of commercial biotechnology companies that offer a variety of EST clones, but these can be expensive.

EST sequence data provide a rapid mechanism for obtaining cDNA sequence data from your gene without the need to screen cDNA libraries. In some cases you may wish to use the EST sequence data for rapid cloning of the target gene by RT-PCR, cDNA library PCR or genomic PCR. This is achieved by designing an oligonucleotide primer complementary to part of the EST sequence for use in conjunction with a 5'- or 3'-gene-specific primer, an adaptor primer or a universal vector-specific primer. The latter is used either for amplification from an existing cDNA library or where the cDNA has been ligated to a vector as a convenient mechanism for adding a universal primer site. If both 5'- and 3'-ESTs are available then two primers could be designed to amplify a selected part of the cDNA clone, such as the protein-coding region.

10.3 Rapid amplification of cDNA ends (RACE)

RACE is a procedure for amplification of cDNA regions corresponding to the 5'- or 3'-end of the mRNA (2) and it has been used successfully to isolate rare transcripts. The gene-specific primer may be derived from sequence data from a partial cDNA, genomic exon or peptide.

3'-RACE

In 3'-RACE the polyA tail of mRNA molecules is exploited as a priming site for PCR amplification. mRNAs are converted into cDNA using reverse transcriptase and an oligo-dT primer as described in *Protocol 8.1*. The generated cDNA can then be directly PCR amplified using a gene-specific primer and a primer that anneals to the polyA region.

5'-RACE

The same principle as above applies but there is of course no polyA tail (*Figure 10.4*). First-strand cDNA synthesis extends from an antisense primer, which anneals to a known region at the 5'-end of the mRNA. However, there is no known priming site available for the subsequent PCR amplification. The trick is to add a known sequence to the 3'-end of the first-strand cDNA molecule as described in *Protocol 10.1*. Terminal transferase, a template-independent polymerase, will catalyse the addition of a homopolymeric tail, such as poly-dC, to the 3'-end of each cDNA molecule. PCR amplification can now be performed using a nested internal antisense primer together with an oligo-dG primer. This will allow the specific amplification of unknown 5'-ends of the mRNA molecule. Alternatively, as discussed for cDNA library construction (Section 10.1), double-strand cDNA synthesis can be followed by blunt ending and adaptor ligation. This provides a specific primer site that in combination with the nested gene-specific primer will lead to amplification of the 5'-end of the cDNA. A common problem with these approaches is that the cDNAs are not always full-length.

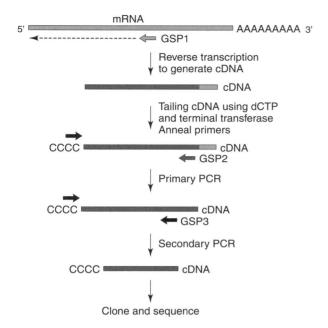

Figure 10.4

Outline of the 5'-RACE technique. Total RNA or mRNA is subjected to reverse transcription using a gene-specific primer (GSP1) priming in the 5' direction. The resulting cDNA is tailed followed by amplification using a tail-specific primer and a nested gene-specific primer (GSP2). Following this a nested amplification reaction is performed using a tail-specific primer and a nested gene-specific primer (GSP3).

A significant advance in the production of full-length 5'-end RACE products is the use of the CapSwitch primer (Clontech). As described in Chapter 8 this allows the addition of a specific primer sequence to the 5'-end of each cDNA by virtue of the homopolymer C-tail added by the reverse transcriptase. This new primer site can be used together with a gene-specific primer for efficient 5'-RACE.

5'- and 3'-RACE

An efficient procedure for cloning both 5'- and 3'-ends of cDNAs or full-length molecules uses adaptor ligation and allows the isolation of both 5'-and 3'-cDNA ends from the same cDNA preparation (3). The adaptor utilizes a vectorette feature for selective amplification of a desired end (Section 10.6) as well as suppression PCR to reduce background amplification (Section 10.1.1).

The technical details of the RACE reaction itself will not be described here since a variety of commercial kits for RACE are available and have optimized protocols and reagents that work very efficiently. These are relatively expensive but more time and money may be spent in optimizing the procedure using a series of independent reagents.

An improvement to standard RACE techniques has recently been reported (4). PEETA (Primer extension, Electrophoresis, Elution, Tailing, Amplification) involves resolving the extension product after reverse transcriptase followed by elution from a gel, then dC-tailing and PCR amplification. It is claimed to be more efficient than the standard RACE procedure and aids in the mapping and cloning of alternatively spliced genes.

Clearly during the design of 5'- and 3'-RACE experiments the primer positions can be located so that the final products have a region of overlap. It is then a simple process to join the two parts of the cDNA by SOEing (Chapter 7). This involves mixing the fragments and performing at least one cycle of PCR, although more cycles can be performed and flanking primers used in the RACE amplifications can be included to amplify the full-length product.

B   Isolation of unknown DNA sequences

It is often of interest to isolate and clone unknown DNA fragments that lie adjacent to already cloned regions of DNA. One obvious example is the isolation of downstream or upstream regulatory regions, including promoters. A further application that is increasingly common is the isolation of flanking regions next to transposon insertions as part of gene knockout strategies. Various approaches to the PCR cloning of unknown DNA sequences will be outlined.

10.4   Inverse polymerase chain reaction (IPCR)

PCR allows the specific amplification of genomic DNA regions that lie between two primer sites facing one another. What if the region of interest lies either 5' or 3' in relation to the primer sites? The answer is inverse PCR (IPCR) (5). The principle of IPCR is shown in *Figure 10.5* and involves the digestion of genomic DNA with appropriate restriction endonucleases, intramolecular ligation to circularize the DNA fragments and PCR amplification. PCR uses primer pairs that originally pointed away from each other but which after ligation will prime towards one another around the circular DNA.

The principle and the protocol for IPCR (*Protocol 10.2*) are the same whatever the application and so as an example the use of IPCR for the isolation of flanking DNA sequences that lie next to a transposon insertion will be described.

Isolation of genomic DNA, digestion and ligation

The success of IPCR is largely dependent on the efficiency of intramolecular ligation of the target DNA fragments within a complex mixture of non-target fragments. A prerequisite is the use of high-quality genomic DNA that should ideally be prepared by using an available commercial kit. The integrity of the DNA should be checked by agarose gel electrophoresis and

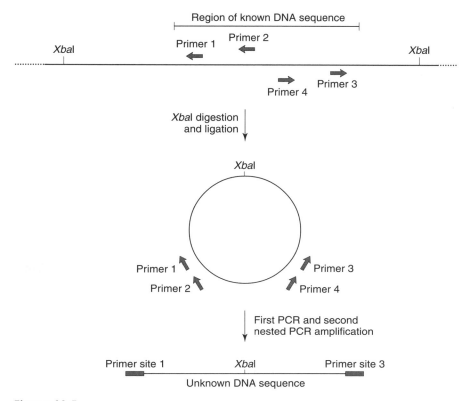

Figure 10.5

Schematic diagram showing the principle of IPCR from genomic DNA. After restriction endonuclease digestion and religation the first-round PCR is performed, in this case using primers 2 and 4. Following this the second-round nested PCR is carried out using primers 1 and 3 which should give rise to one specific amplification product.

should not show any smearing or small molecular size species, including RNA.

A 500 ng aliquot of genomic DNA should be digested with a restriction endonuclease enzyme that digests within the known DNA region, in this case within the transposon, and which will also cut within the unknown DNA region (*Figure 10.5*). It is advisable to set up several different restriction enzyme digests, if possible, since the efficiency of the subsequent PCR amplification decreases rapidly for fragment sizes above 2 kbp in size. Following heating to 70°C to inactivate the restriction enzyme, an aliquot can be retained for gel analysis (see below) and the remainder of the restriction digest reaction should be diluted five-fold in ligation mixture (ligation buffer, H_2O, ligase) and incubated for 6–12 hours at room temperature.

To check the efficiency of restriction digestion and ligation, Southern blot analysis can be performed, in this case using part of the transposon as a probe. An aliquot of the genomic digest should be analyzed along with the ligation reaction. If both the restriction digest and ligation were successful, one hybridizing band should be observed in the genomic digest lane whilst in the 'ligation' lane one hybridizing band of decreased mobility should be visible, due to the circular nature of the ligated product. However, two hybridizing bands are often observed in the ligation sample due to incomplete ligation, as shown in *Figure 10.6*.

First-round PCR

It is important to realize that the first-round PCR is not straightforward, due to the highly complex nature of the template. The reaction is equivalent to amplification of a single copy gene from genomic DNA, but where only a subset of the templates are available for amplification, due to incomplete ligation of the digested DNA. With this in mind, care should be taken when performing the first-round PCR amplification. As described in *Protocol 10.2*, a titration series of the ligation reaction should be used for the first-round amplification in order to maximize the chances of success. Using the outermost primers, a standard PCR amplification should be performed under high-stringency conditions (55–60°C annealing) using a relatively long extension time (2 min) and allowing the reaction to proceed for 40 cycles. The use of 40 cycles ensures that even extremely rare templates are subjected to amplification. A proofreading DNA polymerase should be used to minimize the error rate.

It is useful to analyze an aliquot of the first-round PCR by gel electrophoresis before proceeding to the second-round nested PCR amplification. You may be very lucky and have a single amplification product and in this case you may wish to proceed directly to cloning and sequence analysis to confirm the identity of the product. Generally, however, the outcome is a multitude of relatively weak DNA products, which may or may not be identical in the different restriction digest reactions, but in any case do not provide any indication of the success or failure of IPCR. A second outcome is that no amplification products are detected after the first round of amplification, although again this does not mean that the amplification has failed. The worst outcome is a smear. If heavy smearing appears after

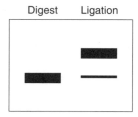

Figure 10.6

Schematic diagram showing a typical Southern blot of digested genomic DNA before and after ligation as part of IPCR. The 'Digest' lane shows detection of a specific restriction fragment corresponding to the target DNA. The 'Ligation' lane shows detection of a larger fragment due to recircularization of the target fragment and also a proportion of DNA that has not ligated and so migrates at the position of the original digested DNA.

the first-round amplification it is highly likely that the second-round nested PCR amplification will fail. Smearing indicates a high degree of nonspecific amplification resulting from either too much template or unsuccessful restriction digestion and ligation.

Second-round nested PCR

The second-round PCR should be viewed as a way of 'fishing' out the specific first-round amplification product from the background of non-specific amplification products. As for the first-round PCR, a titration series should be used, as described in *Protocol 10.2*. This ensures that specific amplification has the best chance of proceeding and avoids smearing due to template saturation. The second-round PCR should be performed with a nested primer pair, at a high annealing temperature using an extension time of 1 min for 35 cycles. Excessive cycling is not required since the amplification will be much more specific, since the complexity of the template is significantly lower than for PCR1. Again a proofreading DNA polymerase should be used.

A single strong amplification product should be observed by agarose gel analysis. Sometimes, however, two or three bands are observed, in which case they should all be cloned and subjected to DNA sequence analysis. This should reveal the specific DNA fragment. If smearing occurs the amount of input DNA should be reduced and the second-round amplification repeated. An example of the result from an IPCR experiment is shown in *Figure 10.7*.

10.5   Multiplex restriction site PCR (mrPCR)

Although IPCR is a relatively rapid way of isolating unknown DNA sequences adjacent to a known piece of DNA, it still requires several time-

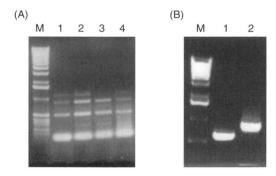

Figure 10.7

Agarose gel showing the primary and secondary PCR amplification products from a typical IPCR experiment. (A) A typical amplification profile from the primary PCR; lanes 1 and 2 represent amplification from one transposon-tagged transgenic *Arabidopsis* line whilst lanes 3 and 4 represent amplification from a second transposon-tagged transgenic *Arabidopsis* line. (B) Results from the secondary PCR amplification; lane 1 represents amplification from primary PCR 1 and lane 2 represents amplification from primary PCR 3.

consuming steps. Multiplex restriction site PCR (mrPCR) eliminates these steps (6) by using a set of sequence-specific primers in conjunction with a set of universal primers that have 3'-sequences corresponding to restriction enzyme sites. Products of mrPCR are analyzed by direct automated DNA sequencing, which means that the whole procedure can be performed in two tubes; one for the first-round PCR and the second for the nested PCR.

Two overlapping primers should be designed from the region of known DNA sequence so that nested PCR can be performed. In addition, four universal primers should be designed that have 3'-sequences matching common restriction sites. Any restriction sites can be used, but for maximum success common six-base recognition site enzymes such as *Eco*RI, *Bam*H1 and *Xba*I are recommended. In some cases six-base enzymes give little success due to the rare distribution of such sites, in which case a four-base recognition site enzyme, such as *Sau*3A, should be used. For the first-round PCR the outermost sequence-specific primer (*Figure 10.8*; SP1) should be used together with all four universal primers (*Figure 10.8*; UP1–4). A 5-fold excess of each universal primer should be used compared with the specific primer. So, in a 50 μl reaction use 50 pmol of the specific primer and 250 pmol of each universal primer. The PCR should be performed as for a standard amplification reaction; however, an extended annealing time of 2 min is recommended and in case of long amplification products, a 3–4 min extension time should be used for a total of 40 cycles. For the second-round nested PCR, 1–10 μl of the first-round PCR should be used together with the 'nested' specific primer (*Figure 10.8*; SP2) and the four universal primers (*Figure 10.8*; UP1–4). After agarose gel electrophoresis one product should appear, although this is not always the case. If two or three amplification products are present they should all be gel purified (Chapter 6) and subjected to DNA sequencing (Chapter 5). Even if only one amplification product is present, it is best to gel purify it prior to DNA sequencing. Once the DNA sequence has been determined and the identity of an amplification product has been verified, the remaining purified DNA fragment should be cloned (Chapter 6) for further analysis.

10.6 Vectorette and splinkerette PCR

Vectorette PCR, also called bubble PCR, was first described by Riley and colleagues (7) as a method for determination of yeast artificial chromosome (YAC) insert–vector junctions. Vectorette PCR provides a method for uni-

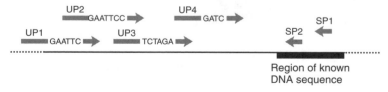

Figure 10.8

Multiplex restriction site PCR. Sequence-specific nested primers SP1 and SP2 are used in combination with various general primers that carry 3'-terminal sequences corresponding to restriction enzyme sites (UP1–4).

directional PCR amplification and is useful for a number of applications including genome walking, analysis of gene structure, promoter cloning and sequencing of YAC and BAC (bacterial artificial chromosome) clones. Vectorette PCR is based on the digestion of DNA and the addition of specially designed adaptors to the digested ends, followed by PCR amplification using a linker-specific primer and a sequence-specific primer (*Figure 10.9*). The adaptors are not completely complementary over their entire

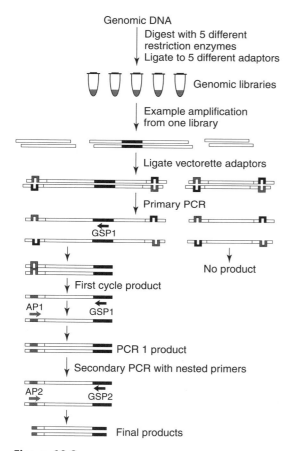

Figure 10.9

Principle of vectorette or genomic walking techniques. Total genomic DNA is digested with different restriction endonucleases followed by ligation to adaptors of known DNA sequence. The generated genomic libraries are then subjected to amplification using an adaptor primer (AP1) and a gene-specific primer (GSP1). AP1 is specific to only one strand of the vectorette adaptor and therefore can only copy DNA after the first cycle of PCR when the GSP1 copies the vectorette adaptor strand to yield a sequence to which the AP1 can anneal. In cases where there is no GP1-directed DNA synthesis no sequence complementary to AP1 is formed and therefore AP1 cannot amplify any DNA. Following the primary PCR amplification a nested amplification is performed using a nested adaptor primer (AP2) and a nested gene-specific primer (GSP2).

length. There is a central region of sequence difference that leads to an unpaired bubble in the adaptor as shown in the example in *Figure 10.10(A)*. This means that there is no sequence in the adaptor that is complementary to the adaptor-specific primer. Such a complementary sequence is only formed during the first cycle of PCR when the gene-specific primer is extended and copies the asymmetric adaptor strand (*Figure 10.9*). The purpose of such asymmetry in the adaptor is to prevent amplification from molecules that do not contain the target sequence. For the isolation of genomic regions next to a known DNA sequence, between 2 and 5 µg of genomic DNA should be digested with various blunt-end producing restriction endonucleases. Between three and five different 'blunt-cutters' should be tested in order to ensure that the length of the subsequent fragment is within the range of efficient PCR amplification. The vectorette adaptors should then be ligated to the ends of the digested DNA, in effect generating a number of genomic vectorette libraries. It is now possible to perform PCR amplification from these libraries using sequence-specific primers and adaptor-specific primers. Two primers should be designed for both the known DNA sequence and the adaptors so that a second-round nested amplification reaction can be performed in order to increase the specificity of the overall reaction to ensure that only correct nested products are amplified. The PCR reactions should be carried out as for standard PCR amplifications; however, it is recommended that an extension time of 2–3 min is employed to ensure amplification of large genomic fragments. Also, commercially available long-range thermostable DNA polymerase mixtures (Chapter 3) often give better results. Vectorette II PCR reagents are available from Sigma Genosys.

The same principle applies to the Universal GenomeWalker™ kit (Clontech), although in this case the adaptors do not contain a bubble, but one strand is short and terminated by an amine group at the 3'-end that cannot be extended (*Figure 10.10(B)*). Thus the complement to the long adaptor strand, which corresponds to the primer and nested primer sites, can only be generated by DNA synthesis from the gene-specific primer.

One problem with standard vectorette technology is that many genes are part of gene families that share considerable DNA sequence identity. Since vectorette PCR is based on the use of a sequence-specific primer together with a vectorette-specific primer one often observes coamplification of other members of the gene family when sequence identity is sufficiently high. One way to overcome this problem, when dealing with human genome analysis at least, is to make use of vectorette libraries constructed from, for example, mouse cell lines that only contain a single human chromosome in the mouse background. This technique was initially developed by Moynihan *et al.* (8), where specific amplification of individual genes from the gene family responsible for early onset of Alzheimer's disease was successful achieved. By constructing vectorette libraries from two different mouse cell lines, GM10479 and GM13139, containing chromosome 14 and chromosome 1 respectively in a mouse background, it was possible using standard vectorette PCR to amplify the individual genes. This technique could also be used for other organisms such as plants where chromosome- or region-specific vectorette libraries can be made, for example, from introgressed lines.

(A)

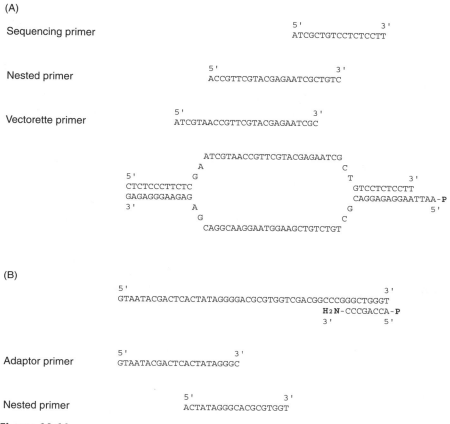

Sequencing primer

5'   3'
ATCGCTGTCCTCTCCTT

Nested primer

5'   3'
ACCGTTCGTACGAGAATCGCTGTC

Vectorette primer

5'   3'
ATCGTAACCGTTCGTACGAGAATCGC

```
                 ATCGTAACCGTTCGTACGAGAATCG
               A                            C
5'           G                            T          3'
CTCTCCCTTCTC                            GTCCTCTCCTT
GAGAGGGAAGAG                            CAGGAGAGGAATTAA-P
3'          A                            G          5'
             G                          C
         CAGGCAAGGAATGGAAGCTGTCTGT
```

(B)

```
5'                                                               3'
GTAATACGACTCACTATAGGGGACGCGTGGTCGACGGCCCGGGCTGGGT
                              H2N-CCCGACCA-P
                              3'          5'
```

Adaptor primer

5'   3'
GTAATACGACTCACTATAGGGC

Nested primer

5'   3'
ACTATAGGGCACGCGTGGT

Figure 10.10

Adaptor sequences. (A) Example of a vectorette adaptor showing the 'bubble' created by the asymmetry of the central region of the adaptor, the adaptor-specific primer and nested primer as well as a primer suitable for direct sequencing of the amplified product. (B) Example of a GenomeWalker™ adaptor and adaptor primers. The lower strand 3'-end is blocked from extension by the amine group and so the adaptor primer binding site is only created once a gene-specific primer has led to copying of the upper adaptor strand. On non-target molecules, since there is no gene-specific primer, the adaptor is not copied and so AP1 has no priming site.

Vectorette PCR can also be used for the isolation of cDNAs. However, instead of digesting genomic DNA followed by adaptor ligation, a cDNA library is constructed followed by ligation of adaptors for the subsequent PCR amplification.

Although generally a specific technique, vectorette PCR can result in unwanted nonspecific amplification involving free cohesive ends of non-ligated free vectorettes and 5'-overhangs of unknown DNA fragments. This is because during the first cycle of the PCR these ends are filled in by the DNA polymerase and during the subsequent denaturation step these ends are able to anneal. The complementary strand of the vectorette primer can then be generated from the unwanted DNA fragments, which ultimately decreases the specificity of the vectorette PCR. To avoid this problem splinkerettes can be used in place of vectorettes. Splinkerettes do not have a central mismatch but have mismatches engineered within them so that they form loop-backed hairpin structures. This reduces the occurrence of end-repair priming and inhibits the DNA polymerase from nonspecific priming at the end of the vectorette. In addition, because small unwanted DNA fragments are more likely to be amplified in the PCR, splinkerettes are better suited for the amplification of longer DNA fragments.

10.7 Degenerate primers based on peptide sequence

The techniques described so far in this chapter, involved in the isolation of unknown DNA sequences, have relied on the presence of known DNA sequence adjacent to the unknown region to be isolated. Sometimes the only available sequence data for a target gene are peptide sequence data derived from the protein encoded by the target gene. In such cases, a common way to isolate a gene of completely unknown DNA sequence from either cDNA or genomic DNA is to use PCR with degenerate primers designed on the basis of such protein sequence data (9,10). There are two principle approaches to the design of degenerate primers: (i) using peptide sequence data obtained from a purified protein; and (ii) using consensus protein sequence data from alignments of gene families.

The purification of proteins can both be time-consuming and also often results in very small amounts of pure protein. However, it is often possible to purify sufficient protein for proteolytic cleavage and gel separation with subsequent transfer to a membrane and solid-phase N-terminal amino acid sequencing of the peptide fragments. Degenerate primers can then be designed by back-translation of the peptide sequence into all possible coding sequences. One of these nucleotide sequences will be the actual sequence that encodes this peptide region in the target gene. The availability of two or more peptide sequences should allow a segment of the target gene to be amplified either from genomic DNA or cDNA. If it is possible to obtain N-terminal sequence data from the mature protein then the orientation of the PCR primer is known. Often, however, mature proteins have been post-translationally modified and do not yield useful sequence data. In this case it may not be possible to order the peptide sequence data relative to the mature protein. This means that the orientations of the PCR primers are not certain and may require the design of two primers from each region, one directed towards the 5'-end and one to the 3'-end. In this way various combinations of primers can be used to ensure that one will represent the correct combination and orientation of a primer pair (*Figure 10.11*). In many experiments involving degenerate primers the length of the expected product cannot be predicted. This is particularly true for genomic amplifications where introns may lie between the peptide segments encoded by different exons. It may be possible from the limited peptide sequence information to identify a homologue in a sequence database which will provide useful information on the relative locations of peptide segments to facilitate rational primer design, and perhaps some indication of the expected size of the product from cDNA amplification.

(A)

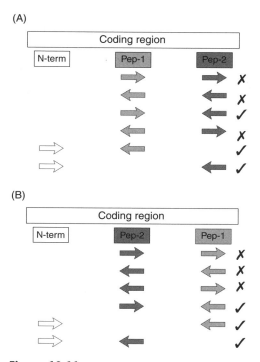

(B)

Figure 10.11

Primer design from peptide sequence data. With the exception of the N-terminus of the protein it is often not possible to assign relative positions for internal peptide sequences generated from a protein-sequencing experiment. So peptides 1 and 2 (Pep-1 and Pep-2) may occur in two possible relative orientations, shown in (A) and (B). If the N-terminal sequence is available then only the reverse primers are required for Pep-1 and Pep-2 regions to generate PCR products. However, if the N-terminus is not known then usually both forward and reverse primers need to be designed for each peptide region and used in various PCR combinations. (A) The pattern of products that could be generated by various primer combinations if the order of peptides is N-term, Pep-1, Pep-2. (B) Products from the order N-term, Pep-2, Pep-1.

If orthologues of the gene of interest have been cloned from other organisms, or if the gene is a member of a gene family, it will be possible to generate protein sequence alignments. These may reveal appropriate regions for the design of degenerate primers, for example, from consensus sequences of highly conserved regions.

Primer design

The first step is to back-translate from the protein sequence to DNA coding sequences using the genetic code. Most amino acids are encoded by more than one codon. In some cases it is possible to select the most likely codon(s) based on the bias in codon usage displayed by the organism. However, the safest approach is to design primers that reflect all possible codons for each amino acid (*Figure 10.12*). Where possible it is advantageous to select protein regions that have amino acids with the fewest possible codon options (M and W [one codon]; C, D, E, F, H, K, N, Q and Y [two codons]) and to avoid regions containing the amino acids L, R, and S which each have six possible codons. Degenerate primers can be produced by 'mixed-base' synthesis where all four nucleotide monomers are added such that 25% of the oligonucleotides contain dA, 25% dC, 25% dG and 25% dT at the degenerate position. An alternative approach is to include a universal base (Chapter 3). As described in Chapter 3 it is important that the 3'-end of the oligonucleotide is perfectly matched to the template, avoiding either a universal or mixed base at the 3'-end. In fact, as shown in *Figures 3.5* and *10.12* it is usually possible to have at least two perfectly matched nucleotides at the 3'-end as these can correspond to codon positions 1 and 2, which are the same irrespective of the third nucleotide of the codon, which show variation. Thus in *Figure 3.5* the 3'-end of the primer is GG, the positions 1 and 2 of a glycine codon. Of course this is another reason to avoid Ser, Leu and Arg which also show variation in codon positions 1 and 2. Even where there are only limited peptide sequence data it is also usually possible to design a nested PCR primer. The 3'-end of the PCR primer is the most critical for determining specificity of

(A) Amino acid sequence
 DNA sequences

	A	D	T	E	W	D	K	G	E	H	G	
NNN	GCA	GAC	ACA	GAA	TGG	GAC	AAA	GGA	GAA	CAC	GGA	NNN
	G	T	G	G		T	G	G	G	T	G	
	C		C						C		C	
	T		T						T		T	

(B) Primer for PCR 1
 (256 sequences)

	GCA	GAC	ACA	GAA	TGG	GAC	AAA	GG
5'	G	T	G	G		T	G	3'
	C		C					
	T		T					

(C) Primer for PCR 1
 with inosine
 (16 sequences)

	GCI	GAC	ACI	GAA	TGG	GAC	AAA	GG
5'		T		G		T	G	3'

(D) Primer for nested PCR
 (32 sequences)

	GAA	TGG	GAC	AAA	GGI	GAA	CAC	G
5'	G		T	G		G	T	3'

Figure 10.12

Example of the design of degenerate primers from amino acid sequence data. (A) The amino acid sequence can be back-translated into all possible codon combinations that might encode the region of peptide sequence. (B) The primer for PCR represents a combination of 256 different sequences; however (C) by using inosine as a universal base for positions of four-base degeneracy the complexity of the mixture can be reduced to 16 different sequences. The primer mix is used together with an appropriate downstream primer in PCR1. (D) Due to the limited amount of amino acid sequence data available, the nested primer overlaps with part of the PCR1 primer, but has been extended so that the 3'-end is different. The use of a single inosine again reduces the complexity of the oligonucleotide mixture to 32 different sequences in this example.

PCR amplification, and so extending the original primer by even two or three nucleotides at the 3′-end should be sufficient to allow amplification of the specific target but not of any nonspecific products formed during the first PCR. The differential specificity of the 3′-end should impose high specificity even though the nested primers overlap significantly with the original primers.

The PCR amplification

The PCR amplification should be performed as for standard PCR, except that usually a higher concentration of primers (100 pmole) is included as only a small proportion of the primers will exactly match the template. In fact under the conditions of a degenerate primer experiment there is some flexibility, in that mismatched primers will also participate in the amplification process, yielding target products that have variation within the regions corresponding to the primer sites. Amplifications using degenerate primers can sometimes be problematic and may require optimization (see Chapter 4). The first parameter to vary is annealing temperature. It is important to keep the annealing temperature as high as possible to avoid extensive nonspecific amplification and a good rule of thumb is to use 55°C as a starting temperature. Ideally if you have access to a gradient block thermal cycler (Chapter 3) set a gradient starting from 55°C up to 65°C using 1°C increments. Alternatively, the magnesium concentration can be titrated from 1 mM to 5 mM in 0.5 mM steps (see Chapter 4). It can often be useful to sample the reaction, every 5 cycles from 25 to 40 cycles, for agarose gel analysis, to allow comparison of relative band patterns and intensities that may indicate the accumulation of a specific product. It is best to use a nonproofreading enzyme for PCR with degenerate primers, as this will prevent exonuclease digestion of the 3′-end of the primer, which would reduce specificity. Normally experiments with degenerate primers are used to provide DNA sequence data for subsequent specific amplifications, which can use proofreading enzymes, or to provide a hybridization probe.

Once you are satisfied that the optimal PCR conditions have been reached it is a good idea to perform nested PCR of the primary reaction using a second set of degenerate primers. This most often leads to increased stringency and the amplification of the correct unknown gene.

Further reading

McPherson MJ, Jones KM, Gurr S-J (1991) PCR with highly degenerate primers. In McPherson MJ, Taylor GR, Quirke P (eds) *PCR1: A Practical Approach*, pp. 171–186. Oxford University Press, Oxford, UK.

References

1. Lukyanov K, Diatchenko L, Chenchik A, Nanisetti A, Siebert P, Usman N, Matz M, Lukyanov S (1997) Construction of cDNA libraries from small amounts of total RNA using the suppression PCR effect. *Biochem Biophys Res Commun* **230**: 285–288.

2. Frohman MA, Dush MK, Martin GR (1988) Rapid production of full-length cDNAs from rare transcripts: amplification using a single gene-specific oligonucleotide primer. *Proc Natl Acad Sci USA* **85**: 8998–9002.
3. Chenchik A, Diachenko L, Moqadam F, Tarabykin V, Lukyanov S, Siebert PD (1996) Full-length cDNA cloning and determination of mRNA 5'- and 3'-ends by amplification of adaptor-ligated cDNA. *BioTechniques* **21**: 526–534.
4. Flouriot G, Brand H, Gannon F (1999) Identification of differentially expressed 5'-end mRNA variants by an improved RACE technique (PEETA). *Nucleic Acids Res* **27**: e8.
5. Ochman H, Gerber AS, Hartl DL (1988) Genetic applications of an inverse polymerase chain reaction. *Genetics* **120**: 621–623.
6. Weber KL, Bolander ME, Sarkab G (1998) Rapid acquisition of unknown DNA sequence adjacent to a known segment by multiplex restriction site PCR. *BioTechniques* **25**: 415–419.
7. Riley J, Butler R, Ogilvie D, Finniear R, Jenner D, Powell S, Anand R, Smith JC, Markham AF (1990) A novel, rapid method for the isolation of terminal sequences from yeast artificial chromosome (YAC) clones. *Nucleic Acids Res* **18**: 2887–2890.
8. Moynihan TP, Markham AF, Robinson PA (1996) Genomic analysis of human multigene families using chromosome-specific vectorette PCR. *Nucleic Acids Res* **24**: 4094–4095.
9. Lee CC, Wu XW, Gibbs RA, Cook RG, Muzny DM, Caskey CT (1988) Generation of cDNA probes directed by amino acid sequence: cloning of urate oxidase. *Science* **239**: 1288–1291.
10. Knoth K, Roberds S, Poteet C, Tamkun M (1988) Highly degenerate, inosine-containing primers specifically amplify rare cDNA using the polymerase chain reaction. *Nucleic Acids Res* **16**: 10932.

Protocol 10.1 5′-RACE

EQUIPMENT

Microcentrifuge

Adjustable heating block or water bath

Thermal cycler

Gel electrophoresis tank

MATERIALS AND REAGENTS

RNA isolation kit (e.g. Qiagen RNeasy® minikit)

DEPC-treated water

Gene-specific primer 1–3 (GSP1–3)

Poly-G primer

First-strand cDNA synthesis kit (e.g. ProSTAR First-strand RT-PCR Kit, Stratagene)

or

Reverse transcriptase buffer: (AMV 10 × RT buffer; 25 mM Tris-HCl pH 8.3, 50 mM KCl, 2 mM DTT, 5 mM MgCl$_2$)

10 mM dNTP mix and 2 mM dCTP solution

Reverse transcriptase (e.g. Qiagen)

Terminal transferase buffer: (10 × buffer; 2 M potassium cacodylate, 250 mM Tris-HCl pH 7.2, 15 mM CoCl$_2$, 2.5 µg ml^{-1} BSA)

Terminal transferase (e.g. New England Biolabs)

Thermostable DNA polymerase and accompanying buffer (e.g. *Pwo* DNA polymerase, Boehringer-Mannheim or KOD DNA polymerase).

0.8% agarose: (100 ml; 0.8 g of agarose in 100 ml of 1 × TAE)

PROTOCOL PART A – FIRST-STRAND cDNA SYNTHESIS

1. Add the following to a 1.5 ml microcentrifuge tube (final reaction volume should be 25 µl):
   - 2.5–5 pmoles of GSP1;
   - 1–5 µg of total RNA;
   - DEPC-treated water up to 12.5 µl.

2. Heat samples to 70°C for 10 min to denature the RNA followed by chilling on ice.

3. Centrifuge the tubes for 5 s in a microcentrifuge to collect the entire content.

4. Add the following to each tube:
   - 2.5 μl of 10 × RT-PCR buffer;
   - 1 μl of 10 mM dNTP;
   - 100–200 units of reverse-transcriptase (the amount required depends on which RT enzyme is used);
   - DEPC-treated water up to 25 μl.

5. Incubate the reaction at 42°C for 1 h followed by heating to 70°C for 10 min.

6. Sometimes it is advisable to purify the cDNA before proceeding but this is not strictly necessary.

PROTOCOL PART B – TdT TAILING OF THE cDNA

1. To each tube add the following:
   - 10 μl of 5 × tailing buffer;
   - 2.5 μl of 2 mM dCTP;
   - water up to 49 μl.

2. Heat the reaction to 94°C for 3 min followed by chilling on ice.

3. Add 1 μl of TdT to each tube and incubate for 10 min at 37°C followed by heat inactivation at 65°C for 10 min.

PROTOCOL PART C – PCR OF TAILED cDNA

1. Add the following to a 0.5 or 0.2 ml PCR tube in a preheated heat block at 94°C:
   - 5 μl of 10 × PCR buffer;
   - 2 μl of 10 mM dNTP;
   - 20 μM of GSP2;
   - 20 μM of poly-G primer;
   - 5 μl of tailed cDNA;
   - water up to 49.5 μl.

2. Add 0.5 μl of *Taq* or other thermostable DNA polymerase.

3. Perform a standard PCR amplification.

4. Analyze amplification products on a 0.8% agarose gel.

5. If necessary perform a nested PCR amplification with GSP3 and the oligo-dG primer as described in Chapter 5.

Protocol 10.2 Inverse PCR from plant genomic DNA

EQUIPMENT

Pestle and mortar

Adjustable heating block or water bath

Microcentrifuge

Thermal cycler

Gel electrophoresis tank

MATERIALS AND REAGENTS

Liquid nitrogen

Plant genomic DNA extraction kit (e.g. Nucleon Phytopure extraction kit, Amersham Lifesciences)

Restriction endonucleases and accompanying buffer

Phenol-chloroform mix (e.g. phenol:chloroform:iso-amyl alcohol 25:24:1, GIBCO-BRL)

T4-DNA ligase and accompanying buffer

Gene-specific primers and nested gene-specific primers

Thermostable DNA polymerase and accompanying buffer (e.g. *Pwo* DNA polymerase, Boehringer-Mannheim)

1% agarose: (100 ml; 1 g of agarose in 100 ml of 1 × TAE)

PROTOCOL PART A – EXTRACTION OF PLANT GENOMIC DNA

1. Harvest fresh tissue and snap-freeze using liquid nitrogen.

2. Grind frozen tissue extensively.

3. Follow manufacturers' instructions for the subsequent extraction procedure.

PROTOCOL PART B – RESTRICTION ENDONUCLEASE DIGESTION

1. For each restriction digest use 500 ng of total genomic DNA in a total volume of 30 μl. Check that the genomic DNA has been digested by gel electrophoresis.

2. Heat-inactivate the restriction enzymes according to the manufacturers' instructions. If heat stable, phenol chloroform extract.

3. Dilute the restriction digest five-fold in ligation mixture (ligation buffer, H_2O, ligase) and incubate at room temperature for 6–12 hours.

PROTOCOL PART C – FIRST-ROUND PCR

1. For each restriction digest use 5 µl, and 10 µl of ligation reaction for each PCR in a final volume of 50 µl.

2. Use the innermost primers for the first-round PCR (*Figure 9.5*; primers 2 and 4).

3. Depending on the primer sequences use high stringency for annealing to avoid minimal nonspecific priming.

4. Perform a standard PCR using 2 min extension at 72°C and allow the reaction to proceed for 40 cycles.

5. Run 15 µl of the PCRs on a 1% agarose gel.

PROTOCOL PART D – SECOND-ROUND NESTED PCR

1. Set up three second-round nested PCRs containing 1 µl, 5 µl, and 10 µl of the first-round PCRs.

2. Use nested primers at high stringency (*Figure 9.5*; primers 1 and 3).

3. Perform a standard PCR using 1 min extension at 72°C and allow the reaction to proceed for 35 cycles.

4. Run 20 µl of the PCRs on a 1% agarose gel.

Genome analysis

11

11.1   Introduction

Genome sequencing projects, whether bacterial, yeast, nematode, plant or human, have become an everyday part of scientific life. Scientific journals and even the science section in national newspapers regularly refer to genome sequencing projects and their potential benefits. In medicine, for example, there is development of rapid diagnostic tests and therapies for disease treatment and prevention, including those based on gene therapies. Furthermore, diseases caused by bacterial pathogens are likely to be an area in which genomic research will have a great impact on new drug development. Prospects for enhancing food production, through more rapid molecular breeding of crop plants and animals, are also highlighted as areas of potential benefit from genome research.

The first genome to be sequenced was the 5386 bp of bacteriophage φX174 (1) and in 1995 the first bacterial genome of *Haemophilus influenzae* (2) was sequenced. Almost 60 microbial genomes have now been completely sequenced, along with more than 100 which are still in progress, and there is intense research to identify essential genes that may be targeted for development of new antibiotics.

The first eukaryotic genome sequence was of the yeast *Saccharomyces cerevisiae* that has some 6 000 genes (3,4). Global transcript analyses to investigate changes in patterns of gene expression when yeast cultures are grown under different conditions are now possible. This became possible through the use of microarrays or oligonucleotide chips representing unique sequences of each of the yeast genes. When such microarrays (5) or oligonucleotide chips (6; Affymetrix.com) are hybridized with fluorescently labeled cDNA isolated from two different cultures it is possible to detect by fluorescence which transcripts are present at increased or decreased levels between the two samples. This provides a picture of exactly which genes are affected by the culture conditions. DNA chip technology including both oligonucleotide chips and DNA microarrays will be important in measuring concerted changes in gene expression in many organisms, tissues and cells.

The first multicellular eukaryotic genome sequence was completed for the nematode *Caenorhabditis elegans* in 1998 and the even more complex *Drosophila melanogaster* sequence was completed in 2000. The information from these genomes offers many advantages for scientific study, particularly as they share many common developmental features with higher eukaryotes including humans. It is possible to access the entire *C. elegans* genome through over 3000 publicly available clones and oligonucleotide primers for over 19 000 *C. elegans* genes can be purchased that yield fragments of around 1.3 kbp for use in microarray and other applications (Research Genetics). The first plant genome sequencing project, of

Arabidopsis thaliana, was completed at the end of 2000, providing new insights into plant gene functionality and providing a basis for comparative genomics with crop plants. Many important crop plants, representing the most central staple foods of mankind, are also being sequenced, with the rice genome completed in 2001 whilst the wheat and maize genome are still in progress. A major landmark in biology came in 2003 when the sequence of the human genome was completed after 13 years.

The molecular mapping of genes and genetic traits and the diagnosis of medical conditions that derive from inheritance and/or spontaneous changes in the genetic material will continue to represent a major application for PCR. It is not possible to describe every PCR-based technique used in genome mapping so we will focus on the most common and well-established techniques, including methods for analysis of mutations, for example in medical diagnosis, and fingerprinting approaches for comparative analysis of individual organisms in forensic applications and to provide molecular markers for gene cloning and agricultural breeding strategies.

11.2 Why map genomes?

The best way to understand protein functionality in any biological system is to examine the phenotypic trait caused by a 'gain-of-function' or 'loss-of-function' mutation to address the question 'what is the biological effect of altering the expression of one or several genes?' Mutations, polymorphisms, and sequence variants (MPSV) have a causal involvement in almost all diseases with genetic components and advances in genome mapping as part of disease diagnosis is therefore extremely important. The mapping of genomes serves several purposes:

- by using the expanding number of genome-specific markers, genetic traits can be more accurately followed through generations;
- genes that cause a genetic defect can be mapped and cloned;
- the mapping of genes allows the design of rapid diagnostic screens for genetic disease traits as well as more informative genetic counseling.

While DNA sequencing provides the highest-resolution genetic analysis data it remains relatively expensive and time-consuming, particularly for the investigation of several markers. In order to generate the DNA for sequencing it is necessary first to amplify the corresponding segments of genomic DNA by PCR. However, the resolution and sensitivity of PCR mean that this technique alone can be sufficient to answer the mapping or diagnostic question without the need for detailed sequence analysis. PCR-based medical diagnosis is very sensitive for mutational detection as it is able to detect abnormal transcripts from one cell in every 10^6 cells. There are many PCR-based mutation detection methods that can distinguish between the presence and absence of alleles of a target gene. As a simple illustration, consider a known genetic locus that gives rise to disease when a small part of the locus has been deleted from both copies of the gene in the genome. The most rapid and reliable approach to analyze this locus is a simple PCR assay using primers spanning the unstable region. When this region of DNA is amplified from an individual it would immediately be informative of the presence or absence of a deletion in one or both copies of the gene (*Figure 11.1*). Thus, if parents were concerned about the possibility that they may have an affected child they could be tested and counseled about the probabilities associated with conceiving an affected child. Often genetic analysis is much more complex, with many potential lesions, predominantly point mutations, existing in a gene. Nonetheless this example illustrates why PCR is used extensively in genome analysis programs. It is also possible to analyze multiple regions of a target gene, or multiple genes, by multiplex PCR in which several different primer pairs amplify different target regions simultaneously in a single PCR tube. Provided the products differ in size, or are differentially labeled, they can be distinguished by subsequent analysis to reveal polymorphisms associated with absence or change in size of a product.

It is equally important to understand what features may limit the utility of PCR in genome analysis and medical diagnostics. First, PCR relies on the availability of good genetic markers within the genome that act as informative 'landmarks'. One potential problem is that different markers 'present themselves' at different stages of the disease, sometimes preventing early disease detection. For example mutations in the *ras* oncogene occur at an early stage during tumorigenesis, whilst mutations in the p53 tumor suppressor gene usually occur in invasive tumors. A second limitation can be the lack of mutational hot-spots in the affected gene, which limits the use of PCR for detection of molecular alterations as part of disease progression and requires the use of DNA sequencing to identify sequence changes. For example, mutations in the *APC* gene, which occur in more than 70% of colon adenomas, show a totally random distribution in this 8.5 kbp transcript. Third, although PCR-based medical diagnosis is designed to increase sensitivity of mutational detection, occasionally such analysis may identify changes in single cells or in cell clusters that are not yet clonal or will never progress into cancer.

11.3   Single-strand conformation polymorphism analysis (SSCP)

The nucleotide sequence of the genome of different individuals of the same species is not identical. Nucleotide substitutions in the human genome

Figure 11.1

PCR-based analysis of a genetic locus carrying a small deletion allows one to distinguish between a wild-type homozygote (A); a deletion-carrying homozygote (B); or a heterozygote carrying one wild-type and one deleted copy of the gene (C).

have been estimated to occur every few hundred base pairs. The traditional way to analyze these polymorphisms was by restriction fragment length polymorphism (RFLP) analysis. RFLP analysis has been used very successfully to create a genetic linkage map of the human genome (7) and has given insight into the chromosomal locations of genetic elements that cause hereditary diseases. While RFLP analysis is useful when distinguishing two alleles at a specific chromosomal locus, it has some disadvantages. First, RFLPs can only be detected if the polymorphisms are present within the recognition site of a restriction endonuclease or where deletions or insertions are present in the region detected by the locus-specific probe. Second, it relies upon restriction of total genomic DNA, Southern blotting and detection of the polymorphism by hybridization with a radiolabelled probe, followed by autoradiography. These methods are time-consuming and expensive.

To make polymorphism detection more efficient, Maniatis and coworkers showed that denaturing polyacrylamide gels can resolve single base-pair substitutions from total genomic DNA due to mobility shifts caused by conformational changes of single-stranded fragments (8). It was subsequently shown that nucleotide sequence polymorphisms can be detected by analysis of mobility shifts of single-stranded DNA caused by single base-pair substitutions (9), known as single-strand conformation polymorphism (SSCP) analysis. With the increased demand for high-throughput mutation analysis, PCR was quickly incorporated as a tool for generating fragments of suitable size for SSCP analysis. The following Sections describe how PCR has simplified SSCP analysis allowing rapid detection of unknown mutations, polymorphisms and sequence variants.

The principle

PCR-SSCP analysis is based on two steps. First, the DNA sequence of interest is PCR-amplified and second, the amplified DNA is heat-denatured and size-fractionated by native polyacrylamide gel electrophoresis (native PAGE). After heat denaturation the mobility of single-stranded DNA fragments is size- and sequence-dependent, with single-stranded DNA molecules adopting secondary structure conformations by intramolecular base pairing (*Figure 11.2*). For a given double-stranded fragment there will be two bands identified following SSCP, one corresponding to each of the two original DNA strands. If two fragments differ by as little as a single base pair, the denatured strands are likely to adopt different conformations and therefore to be distinguishable following native PAGE. The difference is identified by a shift in mobility of one or both of the mutant bands relative to the wild-type control strands (*Figure 11.2*). [α^{32}P] nucleotides are often incorporated into the products during the PCR to label the DNA for PCR-SSCP analysis. Alternatively, a sensitive dye, such as SYBR® Green II (Chapter 9) could be used for 'cold' PCR-SSCP analysis. The migration of single-stranded DNA and the conformational changes are influenced by the percentage of acrylamide, the electrophoresis temperature and the ionic strength of the electrophoresis buffer (see below) and it is recommended that appropriately reproducible conditions are determined for a given fragment.

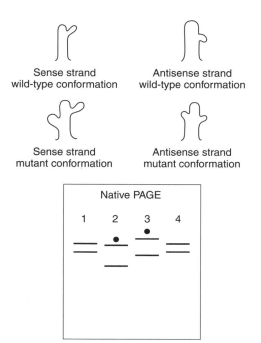

Figure 11.2

Principle of SSCP analysis showing that single base-pair mutation leads to changes in intramolecular conformation of single-stranded DNA fragments which ultimately leads to different electrophoretic mobilities when separated by native PAGE. Mutants are labeled with a solid circle.

The PCR reaction

One advantage of PCR-SSCP is the small quantity of template DNA needed compared with traditional SSCP analysis. Routinely only 5–10 ng of DNA is required for each reaction, compared with SSCP analysis which normally requires between 5 and 10 µg of DNA. An example illustrating this point is the PCR-SSCP based approach to fingerprint sequence variations in ribo-somal DNA of ascaridoid nematodes (10). In this example nematode eggs were collected from the uteri of gravid females by dissection and the isolated genomic DNA was sufficient to perform PCR amplification for PCR-SSCP analysis. Two primers were designed to regions of the 5.8S and the 28S rRNA genes of *C. elegans* and labeled with [γ^{33}P]ATP. Following PCR it is advisable to check a small aliquot of the reaction products (1/25) on a standard 2.5% agarose gel to ensure that the amplification reaction was successful. As discussed below, a dilution series of the amplification reaction should be diluted in SSCP loading buffer (10 mM NaOH, 95% formamide, 0.05% bromophenol blue and 0.05% xylene blue) and denatured at 95°C for 5 min, followed by snap-cooling before being subjected to native PAGE.

Factors affecting the quality of single-stranded DNA mobility

A number of parameters affect the quality and reproducibility of PCR-SSCP analysis. Temperature influences the conformation of single-stranded DNA and should therefore be kept constant. Very often SSCP analysis is performed by running gels at room temperature at low power (~ 10 W). This is acceptable if the room has temperature control such as air conditioning. If not it is better that a constant electrophoresis temperature be maintained by running the nondenaturing gel in the cold room (4°C) at 40–50 W.

A second important factor is the percentage of acrylamide and the degree of cross-linking by bis-acrylamide. Reduced cross-linker concentrations may improve the fragment separation quality. However, this is not always the case and must be tested for the individual fragments. Adding additives to the gel matrix may also have a positive effect on the overall separation quality. For example, addition of 2.5–10% glycerol increases purine-rich fragment mobility compared with pyrimidine-rich strands due to polar-OH group interactions or new intermolecular hydrogen bond formation. This phenomenon is not limited to glycerol but is also seen when sucrose, glucose, formamide and dimethyl sulfoxide are used. It has been proposed that glycerol makes the folded structure of single-stranded DNA more relaxed, acting like a weak denaturing agent. Again optimal glycerol concentrations, to achieve the best separation quality, must be determined empirically.

The ionic strength can affect the resolution quality of the gel. In some cases it is advantageous to decrease the ionic strength of the electrophoresis buffer to 0.5 × TBE (1 1 5 × stock solution: 54 g Tris base, 27.5 g boric acid, 20 ml 0.5M EDTA pH 8.0), although this also lowers the buffering capacity. Another option is to lower the ionic strength of the stacking gel, which will increase the sharpness of the banding pattern.

In some cases single strands anneal to form duplex DNA prior to or during electrophoretic separation and this can affect polymorphism detection. This is more likely to occur if several fragments are analyzed in one lane. To overcome this potential problem the PCR samples should be diluted and the extent of dilution will depend on the amount of PCR product. However, a good starting point is to make dilutions of 1:200, 1:50, 1:10 and 1:2. Overloading the gel increases the possibility of abnormalities in migration patterns.

Traditionally when performing a problematic PCR an increase in primer concentration may be advantageous, but in PCR-SSCP this is not the case. By increasing the primer concentration a dose-dependent decrease in SSCP resolution is often observed due to primer–fragment reannealing. This problem can be overcome by purifying the PCR products from the primers (Chapter 6) prior to electrophoretic separation.

Fragment size for detection of polymorphisms

Using PCR-SSCP for the analysis of mutations in large fragments is difficult. The best results are achieved using fragments around 150 base pairs in length and fragments longer than 200 bp should not be used. If the use of larger fragments cannot be avoided, it is possible to increase the chance of detecting polymorphisms by performing a restriction endonuclease digestion on the fragment, which should allow the simultaneous analysis of the resulting smaller fragments (*Figure 11.3*).

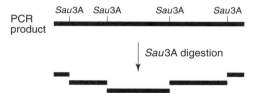

Figure 11.3

Restriction endonuclease digestion of PCR-generated DNA fragment for use in SSCP analysis. A four-base recognition site enzyme, such as *Sau*3A, which will restrict the DNA on average every 256 bp, should be used.

11.4 Denaturing-high-performance liquid chromatography (DHPLC)

The high-throughput discovery and analysis of single nucleotide polymorphisms (SNPs) relies heavily on automated techniques and ideally only requires simple PCR amplification from samples. DHPLC represents such a technique allowing the automated detection of single base pair substitutions and small insertions or deletions. The technique is based on the fact that under partial denaturing conditions DNA heteroduplexes, representing DNA with different sequences, formed after mixing, denaturation and reannealing, will be retained less than the corresponding homoduplex on a DNA separation matrix. The stationary phase, commercially available from Transgenomic Inc. as DNASep™, is made of 2–3 μM alkylated nonporous poly(styrene-divinylbenzene) particles, which allow separation of nucleic acids by means of ion-paired reverse-phased liquid chromatography. By using a hydro-organic eluent containing an amphiphilic ion such as triethylammonium ion and a hydrophilic counter-ion such as acetate, the separation of DNA fragments is achieved. The retention itself is governed by electrostatic interactions between the positive surface potential formed by the triethylammonium ions (or similar amphiphilic ions) at the stationary phase and the negative surface potential generated by the dissociated and exposed phosphodiester groups of the DNA. Because of this, double-stranded DNA is retained according to the length (11). Retained double-stranded DNA is then eluted using an increase in concentration of an organic solvent such as acetonitrile. DHPLC has been used successfully for microsatellite analysis (Section 11.13), human identification and parentage testing, detection of loss of heterozygosity in tumors, DNA methylation analysis and quantification of gene expression (11).

Although DHPLC represents a high-throughput and sensitive method that does not require special reagents or post-PCR treatment, it does require DNA samples at the same quality as for DNA sequencing. However, by carefully designing your PCR (Chapter 4) this should not present a problem. For further information on DHPLC see www.transgenomic.com and for a detailed review of DHPLC and its uses we recommend an article by Xiao and Oefner (11) or the website http://insertion.stanford.edu/pub_method.html.

11.5 Ligase chain reaction (LCR)

Successful detection of single base pair allelic differences is often required in genetic disease detection or forensic DNA analysis. The ligase chain reaction (LCR) allows detection of single base pair substitutions with high specificity and reliability. LCR is not really a PCR method since there is no involvement of a polymerase; rather amplification of the product relies upon the action of a ligase, ideally a thermostable DNA ligase, to join the ends of two adjacent primers. The underlying principle of LCR is the ability of DNA ligase to ligate two adjacent oligonucleotides only if the junction between the two is perfectly base-paired to the target DNA (*Figure 11.4*). This means that a single base-pair substitution at the ligatable end of either primer would prevent ligase forming a covalent linkage.

For increased specificity, the oligonucleotides should be of a sufficient length (>20 nt) to allow hybridization to the unique genomic target region and the ligation reaction should be performed near to the melting temperature (T_m) of the two oligonucleotides. The use of a thermostable DNA ligase allows multiple cycles of LCR to be performed, allowing a linear accumulation of ligated product. The amount of product may be further increased by using both strands of genomic DNA as targets together with two sets of adjacent oligonucleotides complementary to each strand. This in effect means that the product can be increased exponentially by repeated thermal cycling. Target-independent ligations are minimized either by addition of salmon sperm DNA or by using oligonucleotide primers that create single 3'-overhangs.

In contrast to PCR the LCR requires high-quality genomic DNA. The oligonucleotides should be radiolabelled at the 5'-end and used directly in the subsequent LCR; 40 fmol of radiolabeled primers should be added to the genomic DNA followed by addition of thermostable DNA ligase. The reactions should then be incubated at 94°C for 1 min followed by 65°C for 4 min for 20–30 cycles.

The specificity of *T. aquaticus* DNA ligase is very high, showing efficient ligation when correctly base-paired oligonucleotides are used, with near to zero ligation in the presence of a single mismatch. However, some 'incorrect' ligation events (less than 1%) do occur, which can be overcome by reducing the input DNA amount (12).

11.6 Amplification refractory mutation system (ARMS)

The basic principle of the ARMS assay is similar to that of the LCR assay (Section 11.5), which relies on the discrimination of the 3'-nucleotide of primers for different alleles. ARMS is also known as allele-specific PCR (ASP), PCR amplification of specific alleles (PASA) or allele-specific amplification (ASA). In an ARMS assay (*Figure 11.5*) two PCR reactions are set up using the same template DNA, one common primer but with different allele-specific primers (13). In one reaction a primer is used that will amplify from

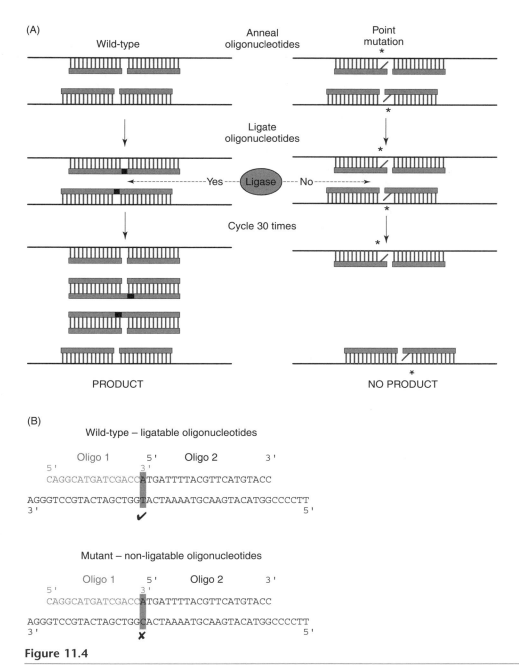

Figure 11.4

Principle of the LCR. (A) Annealing of complementary oligonucleotides to the target DNA occurs at a temperature close to the melting temperature of the resulting duplexes. This allows ligation of perfectly annealed oligonucleotides, but not those in which the upstream oligonucleotide has a 3′-mismatch. Repeated cycles allows exponential accumulation of ligated product. (B) Examples of primers for LCR showing annealing of two complementary primers to a wild-type template yielding a substrate for ligase (✔) and to a mutant template with a single nucleotide mismatch to the 3′-end of the upstream oligonucleotide thus preventing (✘) ligation.

one allele while in the second, parallel reaction a primer is used that will amplify from the other allele. These primers are specific for only one allele because their 3'-nucleotides are different and correspond to a position of nucleotide variation between the alleles. It is critical that *Taq* or a similar DNA polymerase that does not possess 3'→5' proofreading activity is used in PCR (Chapter 3). A proofreading enzyme would correct the mismatched 3'-position on the primer, destroying the discrimination of the assay. In

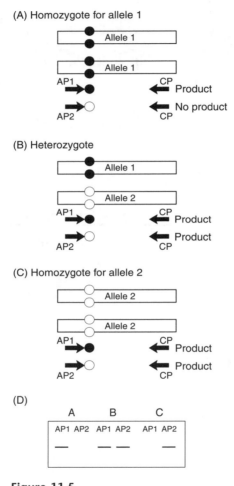

Figure 11.5

ARMS assay for discriminating between alleles. Two allele-specific primers (AP1 and AP2) are designed so that their 3'-ends are able to anneal to regions that differ between the two alleles. In combination with a common primer (CP) these allele-specific primers can amplify a product only if the allele to which they can anneal is present. Thus in (A), for an allele 1 homozygote, only AP1 can amplify the product. The absence of allele 2 prevents AP2 from functioning. In (B), a heterozygote in which both alleles are present, both reactions will yield a product; while in (C), only AP2 can generate a product. (D) shows a schematic gel indicating the pattern of products expected from these different genotypes.

addition the enzyme should not be able to initiate DNA synthesis from a mismatch site so that there is essentially no amplification and therefore no product. This should contrast with efficient amplification and a clear product when the primer is perfectly matched. The products can then be analyzed simply by agarose gel electrophoresis (Chapter 5). Since an expected outcome of the analysis of some DNA samples is the absence of a product, it is important to ensure that the PCR is working efficiently and the simplest way of ensuring this is to co-amplify a different genomic region using control primers.

Primers for an ARMS assay are designed from genomic DNA sequence and should ideally be around 30 nucleotides in length. Discrimination may be enhanced by introducing further mismatches close to the 3'-end of the primer, although if these are not carefully considered, yield may be compromised. The closer to the 3'-end a destabilizing mismatch is placed the more significant the destabilizing effect. Not all mismatches are equally destabilizing, with the following rank order: CC > CT > GG = AA = AC > GT. The effectiveness of any combination of mismatches must be determined empirically for their utility in the ARMS assay being established. The common primer should be of similar length, with about 50% GC content and positioned to generate a conveniently sized ARMS PCR product. The control primers, for checking PCR efficiency, should be designed to have similar properties to the common primer, and these should yield a fragment that is of a distinct size to the ARMS product to allow simple identification by agarose gel analysis. The common and control primers should be checked to ensure that they do not display complementarity to the 3'-ends of either allele-specific primer. The control primers can sometimes influence the efficiency and specificity of the ARMS assay.

ARMS analysis has been used for a large number of polymorphisms, germline and somatic mutations, carrier status determination, prenatal diagnosis of inherited disease and detection of residual disease during and after cancer therapy.

11.7 Cleaved amplified polymorphic sequence analysis (CAPS)

A well-established, but time-consuming method for detecting single nucleotide polymorphisms (SNPs) is RFLP analysis. A PCR-based alternative that has been widely used in the analysis and mapping of plant genomes is CAPS. The method relies on the fact that regions of known genes can be amplified and nucleotide differences between two alleles lead to a difference in a restriction site. Thus the two alleles can be distinguished by PCR amplification followed by appropriate restriction endonuclease digestion and analysis of the products. In one case the product will be cleaved by the restriction enzyme while in the other it will not. This approach was developed by Konieczny and Ausubel (14) for 18 loci in *Arabidopsis thaliana* to discriminate between the ecotype Landsberg erecta and Columbia. Each CAPS marker could be discriminated by at least one restriction enzyme digest which allows efficient mapping of cosegregating genes by direct PCR analysis and restriction digestion. A limitation of RFLP and CAPS analysis is that the SNP must occur within the restriction site for an enzyme, therefore the majority of SNPs will not be detectable. More recently the approach has been modified through the use of modified primers that introduce additional mismatches that create a restriction site only when the SNP occurs, allowing easy discrimination between alleles (15,16). The approach relies upon amplification of relatively short DNA fragments, as the restriction site is being introduced within the primer and therefore in general the size discrimination between the two alleles will only be around 20 or so base pairs. For example, the two alleles shown below can be discriminated by the primer-directed introduction of a G in place of an A leading to the introduction of a *Mbo*I restriction site (15).

The use of the primer to amplify from both alleles:

```
Primer      GGATCTCGCCGAGAACGA
Allele 1  ...GGATCTCGCCGAGAACAACCGTGGAG...
Allele 2  ...GGATCTCGCCGAGAACAATCGTGGAG...
```

leads to PCR products that differ in their ability to be cleaved by *Mbo*I that recognizes the sequence GATC:

```
Allele 1  ...GGATCTCGCCGAGAACGACCGTGGAG...
Allele 2  ...GGATCTCGCCGAGAACGATCGTGGAG...
```

11.8   SNP genotyping using DOP-PCR

SNP genotyping is generally performed by PCR amplifying SNP-containing loci individually. Although multiplexing can increase the speed of SNP genotyping, the difficulty of predicting and determining compatibility as the complexity increases is challenging (Section 11.11). A method has been developed that achieves a broader representation of possible amplifiable sequences by applying degenerate oligonucleotide-primed (DOP)-PCR with SNP genotyping (17). During the amplification step the partially degenerate primers bind to many sites within the genome and amplify a product where two sites lie close to one another in opposite directions. This results in amplification of a mixture of DNA fragments many of which contain SNPs. Following this the SNPs can be subjected to genotyping directly from the DOP-PCR.

DOP-PCR utilizes a partially degenerate primer where about one-third of the positions in the center of the primer are degenerate. It has been shown that by varying the length of the unique region from six to ten nucleotides DOP-PCR results in a genome complexity reduction that supports effective SNP genotyping in species such as human, mouse, and *Arabidopsis thaliana* (17). When the 3′-end of such a degenerate primer consists of six unique nucleotides, a huge proportion of any eukaryotic genome can be amplified, most often resulting in a smear of DOP-PCR products on a denaturing PAGE. However, by increasing the number of unique nucleotides to eight or ten a discrete number of products is generally observed, and in terms of the human genome such primers result in the amplification of a few hundred unique products.

Fragment size is affected by the DOP-PCR cycling parameters and a two-part cycling program should be used. The first five cycles should have a very low annealing temperature followed by 35 cycles at a higher annealing temperature. This ensures primer binding during the initial cycles whereas the later cycles improve specificity and yield. By lowering the annealing temperatures and/or shortening extension times, the size range of the amplified products can also be shifted downward.

A critical factor for the use of a reduced-complexity genome-wide PCR amplification for SNP genotyping is the reproducibility of the PCR. Using the same parameters on 32 human genomic DNA samples on two different thermocyclers, Jordan *et al.* (17) obtained similar banding patterns, demonstrating the robustness and reproducibility of the technique.

DOP-PCR product complexity is the most important variable for direct SNP genotyping. If the complexity is too low, then only a few SNP-containing fragments are amplified. However, as the complexity approaches that of the whole genome, the efficiency and accuracy of SNP genotyping becomes problematic. The effectiveness and accuracy of direct SNP genotyping on DOP-PCR products has been shown by sequencing samples of SNPs from fragments amplified in nine DOP-PCRs from human, mouse, and *Arabidopsis*.(17).

11.9   Random amplified polymorphic DNA (RAPD) PCR

RAPD-PCR is also known as arbitrarily primed PCR (AP-PCR) and is a relatively rapid PCR-based genomic fingerprinting method. However, due to occasional production of nonparental products it is not recommended for use in such applications as paternity testing where unequivocal results are demanded. It provides a very useful tool for genome analysis in bacterial, fungal and plant identification and population studies where individual isolates can be compared rapidly. For example they can be used to identify pathogens or the occurrence of particular strains/pathotypes, and RAPD markers that cosegregate with pathogenesis traits can provide an important tool for identifying pathogenic strains of bacteria and fungi. In plant studies RADP-PCR provides a useful tool for plant breeding programs by providing markers associated with traits to examine trait heritability. The approach is also used for the detection of abnormal DNA sequences in human cancer. Commonly, AP-PCR genomic fingerprints of DNA from normal and tumor tissue can be used to compare deleted or amplified DNA sequences in cancer cells.

AP-PCR uses a single primer to initiate DNA synthesis from regions of a template where the primer matches imperfectly. In order for this to work, the initial cycles have to be performed at low stringency (37–50°C), normally for the first five cycles, which allows hybridization to imperfect sites throughout the genome. The stringency is then increased (55°C) as for standard PCR amplification and the reaction allowed to proceed for an additional 30–35 cycles. In effect this means that only the best mismatches during the initial amplification cycles are further amplified. By careful optimization it is possible to obtain between 50 and 100 distinct DNA fragments which can then be separated by PAGE. Since AP-PCR is based on arbitrary amplification under low stringency conditions, various genomic regions can be amplified simultaneously in a single PCR amplification. AP-PCR can allow visualization of deleted or amplified DNA fragments with different intensities, which in turn allows genomes to be differentiated in a quantitative and qualitative manner. However, although intensity differences may be observed they may not represent real deletions or amplifications but may reflect genetic polymorphisms in the human population.

A key advantage of AP-PCR is that targeted DNA fragments can be re-amplified and cloned using the same primers as used in the initial PCR amplification. However, there are also problems associated with AP-PCR. First, the reproducibility of the banding patterns may vary from day to day even when using the same conditions and the same primers. Second, it has been demonstrated that both MgCl$_2$ and template concentrations can affect the banding pattern. Third, although low stringency conditions are used during the initial cycles, low annealing temperatures may also affect the outcome of the analysis.

11.10 Amplified fragment length polymorphisms (AFLPs)

AFLP displays a subset of PCR products derived from restriction digestion of genomic DNA (18). Genomic DNA is digested with a restriction enzyme and then double-stranded oligonucleotide adapters are ligated to the ends of the fragments. A subset of fragments is then amplified by PCR using primers designed to include part of the adapter, the restriction site and about three nucleotides beyond the site. Since only a proportion of the DNA fragments will display complementarity to these three 3'-nucleotides, only this subset of fragments will be amplified. A nonproofreading enzyme such as *Taq* DNA polymerase must be used to ensure that the specificity of the 3'-end of the primer is maintained. The result is that many fragments will be simultaneously amplified and a radiolabeled dNTP can be included to label the products, which can be separated by electrophoresis through a DNA sequencing gel.

The approach can be used to compare samples that display phenotypic traits, and bands that appear in one phenotype but not the other may represent markers linked to the phenotypic marker. The use of AFLP markers can provide powerful tools for molecular breeding strategies in both plants and animals. The AFLP approach has also been used for comparative analysis of cDNA populations thus allowing differential analysis of expressed genes (19). Detailed protocols for AFLP technology are available at (http://www.dpw.wau.nl/pv/index.htm) under the subheading Documents.

11.11 Multiplex PCR analysis of *Alu* polymorphisms

Alu sequences probably represent the largest family of short interspersed elements and were first defined as renatured repetitive DNA that was distinctively cleaved with the restriction enzyme *Alu*I (20). In the human haploid genome *Alu*s are present in excess of 500 000 copies constituting approximately 5% of the genomic DNA by mass and having an average length of 300 bp. *Alu*s are ancestrally derived from the 7SL RNA gene and move throughout the genome by retrotransposition (21). As genetic markers *Alu*s have several exceptional features including: (i) stability of insertion; (ii) unknown mechanism for removal from their chromosomal location; and (iii) some *Alu*s have not yet reached a stable chromosomal fixation. The low rate of *Alu* loci that reach polymorphic levels dictates that the probability of an independent polymorphic insertion into the same chromosomal site is extremely unlikely. Thus *Alu* insertions have proven extremely useful as genetic markers in DNA fingerprinting, population studies, forensic applications and paternity determinations. Rapid analysis of *Alu* polymorphisms is conveniently achieved using multiplex PCR techniques with multiple primer combinations to amplify more than one *Alu* locus in a single assay.

Duplex and triplex PCR reactions can be more difficult and temperamental than a PCR with a single primer pair. Primers for *Alu* amplification should be designed as complementary pairs directed to the 5' and 3' single copy flanking sequences of each *Alu* insertion. *Alu* sequences are easily accessible through databases such as ALUGENE (http://alugene.tau.ac.il/). Before embarking on duplex or triplex reactions it is important to ensure that the reaction conditions for each primer pair are optimized and give rise to a specific amplification product. Samples in *Alu* analysis are normally amplified in 50 µl reaction volumes using a titration of DNA ranging from 10 ng to 100 ng for each primer combination. Optimal primer concentrations must be determined empirically depending on the strength of amplification of each locus but typically range from 0.05 µM to 0.5 µM. For weakly amplified loci the primer concentration should be increased whilst for 'strong' loci the primer concentration should be decreased. In addition, it is often necessary to have several different thermal cycler programs optimized for the different primer combinations. Apart from these parameters the amplification reactions should be designed as for standard PCRs.

11.12 Variable number tandem repeats (VNTR) in identity testing

PCR highly polymorphic variable number tandem repeats (PCR-VNTRs) are extremely important markers in identity testing and form an integral part of forensic DNA analysis. VNTR loci variants were originally characterized by restriction fragment length polymorphism (RFLP) analysis. However, with the advent of the PCR, VNTR analysis methods have become more reliable both in terms of sensitivity and specificity. VNTRs can be divided into short tandem repeats (STRs) with a fragment length between 100 bp and 300 bp and amplified fragment length polymorphisms (AFLP) with fragment lengths between 350 bp and 1000 bp. Various commercial kits are now available to facilitate such analyses. The following sections describe the use of VNTR/STR analysis in identity testing as part of forensic DNA analysis.

Short tandem repeat (STR) amplification

STR loci are scattered around the human genome and are highly polymorphic in length and often in the sequence of repetitive elements of 3–7 nucleotides. In forensic studies allelic differences of 4–5 bp repeats are most commonly used (*Table11.1*). An advantage of STRs in forensic work is that even partially degraded DNA often provides a template for successful PCR amplification due to the small size of the STR loci. Polymorphism analysis of STR loci is almost exclusively based on polyacrylamide gel electrophoresis (PAGE). Agarose gel electrophoresis can also be employed to give lower-resolution data.

Table 11.1 Commonly used STR loci in forensic studies

STR locus	Chromosomal location	Repeat sequence 5'→3'	Primer sequences 5'→3'
HUMFES/FPS	15q25–qter	AAAT	GGG ATT TCC CTA TGG ATT GG
			GCG AAA GAA TGA GAC TAC AT
HUMF13A1	11p15.5	AATG	GAG GTT GCA CTC CAG CCT TT
			ATG CCA TGC AGA TTA GAA A
HUMCSF1PO	5q33.3-34	AGAT	AAC CTG ACT CTG CCA AGG ACT AGC
			TTC CAC AGA CCA CTG GCC ATC TTC
HUMTPOX	2p23–2pter	AATG	ACT CGC ACA GAA CAG GCA CTT AGG
			GGA GGA ACT GGG AAC CAC ACA GGT

General considerations

PCR allows the amplification of one or more STR loci simultaneously in a multiplex PCR. However, if sensitivity and robustness of analysis is the main issue it is advisable to amplify only one polymorphic STR. Success of multiplex analysis relies on different STR loci being amplified reliably together, which in turn relies on compatible amplification parameters such as similar primer annealing temperatures. The migration pattern of different STRs is also important and, particularly for manual analysis, different STR loci should ideally not yield overlapping mobility patterns. However, fluorescent-labeled primers with different fluorophores are increasingly used to successfully analyze alleles of overlapping sizes using analytical equipment such as an automated DNA sequencer.

PCR amplification

PCR amplification parameters for STR analysis are essentially the same as for a standard PCR (Chapter 2). However, there are some general considerations that should be noted. All procedures should be performed under sterile conditions using gloves, aerosol-resistant pipettes and a laminar airflow sterile bench (Chapter 4). The amount of template DNA is particularly crucial for successful amplification and should be in the range 1–25 ng. It is critical to include a positive control for the amplification and in most forensic DNA analysis cases, DNA from cell line K562 is employed for this purpose, which can be purchased from commercial sources. Another critical consideration is the number of amplification cycles that should be adjusted for each STR locus. It is advisable not to exceed 30 cycles due to the increased chance of nonspecific amplification.

Detection of polymorphic STRs

Perhaps the most important aspect of polymorphic STR analysis is the detection of allelic differences after the PCR amplification itself. Most laboratories now use automated detection of STR profiles using fluorescently labeled primers in conjunction with automated DNA sequencers. In most cases a fluorophore is attached to the 5'-end of one primer, becoming incorporated into the amplified product, which in turn can be detected by using an automated DNA sequencer. If such instruments are not available there are manual detection methods using agarose and PAGE. Both yield good results although with differing resolution. Agarose gels can distinguish amplification products differing by 4 bp when using an agarose concentration between 4 and 5% whilst polyacrylamide gels can accurately distinguish alleles differing by a single base pair. The main advantages of agarose gel electrophoresis are that it is simple and cheap, and amplification products can be rapidly analyzed which in turn speeds up optimization experiments. A further advantage is that the amplification products can be loaded directly onto the gel and subsequently manipulated. However, for more complex STR profiling, such as multiple loci amplification procedures, PAGE is recommended due to the improved resolution of amplification products. PAGE has been described elsewhere in this book (Chapter 8) although it is worth highlighting factors that may affect the overall result. Nondenaturing PAGE conditions have both advantages and disadvantages over denaturing conditions. Nondenaturing conditions allow additional variants of STR profiling to be detected such as differences in hypervariable allelic regions. However, a large number of STRs are A/T rich and hence show electrophoretic anomalies due to helical axis distortion. A second problem with nondenaturing conditions is that intra-allelic differences are most often only detected under very specific conditions, which means that standard conditions must be used in each analyzing laboratory. For maximum reliability and consistency the use of denaturing PAGE is therefore recommended.

Validation studies

Before any new STR system can be used as part of forensic work it should be extensively validated to ensure reliability of the results. The importance of validation studies for forensic DNA profiling cannot be overemphasized. We will describe several important points related to STR validation below whilst further information on STR validation studies can be found at http://www.cstl.nist.gov/biotech/strbase/valid.htm and in Anderson *et al.* (22). In cases involving population studies, a rule of thumb is to examine a minimum of 200 unrelated individuals or 500 meioses for each STR system to determine allele frequencies, Mendelian inheritance and mutation numbers. Furthermore, it is very important that the PCR system employed resolves allelic differences from complex mixtures of material such as vaginal cells/semen, saliva, hair roots and blood. In effect this means that the extraction procedure, the storage conditions for PCR mixtures, and the subsequent PCR amplification must be tested for reliability and consistency between various sample mixtures and individuals.

11.13 Minisatellite repeat analysis

Minisatellite repeat coding as a digital approach to DNA typing was pioneered by Professor Sir Alec Jeffreys (23) and although an 'old' technique in terms of PCR applications the approach deserves recognition and consideration due to its enormous impact on DNA typing.

Minisatellite repeat coding is based on assaying sequence variations in minisatellite alleles rather than length differences. Minisatellite alleles vary not only in the number of repeat copies but also in the interspersion pattern of variant repeat units. The hypervariable locus D1S8 (MS32 probe) shows two classes of repeat units, a-type and t-type, that differ by a single nucleotide substitution which creates/destroys a *Hae*III restriction endonuclease site. Before the development of the minisatellite repeat coding technology, *Hae*III positive and *Hae*III negative repeat units were determined by PCR amplification of the entire allele followed by partial *Hae*III digestion. This approach was limited in that alleles larger than 5 kb were outside the range of efficient PCR amplification. The development of the minisatellite repeat coding technology now allows minisatellite variant repeat (MVR) mapping of MS32 alleles of any length. The technique, outlined in *Figure 11.6*, is based on using two MVR-specific primers that prime from either an a-type or a t-type repeat unit and a primer that is fixed in the minisatellite flanking DNA (32D). The a- or t-type specific primers each carry a 5'-tail called the TAG sequence that becomes incorporated into all the products. Initial PCR amplification occurs at low selective primer concentration (a-type or t-type specific) to generate two ladders of products from the ultravariable end of any MS32 allele. These products terminate in 32D and TAG sequences and so are efficiently amplified by the high concentrations of the 32D and TAG primers in the reaction, generating ladders of PCR products that are detectable by Southern blot analysis. The ladders normally extend beyond 3 kb (100 bp units) into each allele which can then be used for the generation of allele binary codes. It is possible to increase the number of amplification cycles and to visualize the amplification products directly using ethidium bromide; however, over-amplification can also lead to the loss of minisatellite information for the first 10–15 repeat units. The variation in diploid codes is easily detected using minisatellite repeat coding. Jeffreys *et al.* (23) demonstrated this by performing MVR-PCR typing of 334 unrelated individuals revealing an average of 30 code mismatches per pair of individuals within the first 50 repeat units. It was shown that no two individuals had the same MVR code and all individuals could be distinguished using only the first 17 repeat units. MVR-PCR is very simple and also sensitive and has been used extensively in forensic DNA analysis. Profiles can be generated from as little as 10 ng human DNA. Even less DNA can be used although this often gives rise to random fluctuations in band intensities. The technique can also be employed when dealing with degraded DNA because the only requirement for minisatellite repeat typing is the presence of the 32D primer and at least 30 repeat units. MVR-PCR can also be used for mixed samples, particularly if pure DNA is available from one of the individuals. As a result MVR-PCR is often used in rape cases where pure DNA is available from the victim. MVR-PCR analysis procedures have been described extensively elsewhere

(A)

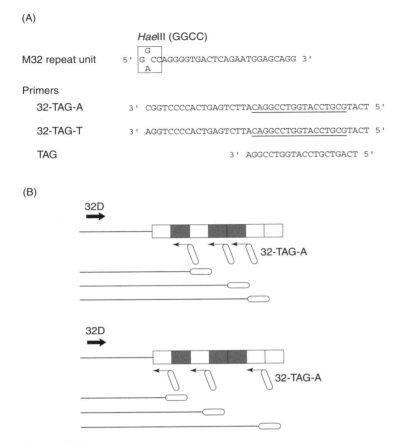

*Hae*III (GGCC)

M32 repeat unit   5' [G / A] G CC AGGGGTGACTCAGAATGGAGCAGG 3'

Primers

32-TAG-A   3' CGGTCCCCACTGAGTCTTACAGGCCTGGTACCTGCGTACT 5'

32-TAG-T   3' AGGTCCCCACTGAGTCTTACAGGCCTGGTACCTGCGTACT 5'

TAG   3' AGGCCTGGTACCTGCTGACT 5'

(B)

Figure 11.6

Minisatellite variant repeat (MVR) mapping. Minisatellite alleles occur in two distinguishable forms, a-type (shaded box) carrying a *Hae*III restriction site and t-type (open box) with a single base change that destroys the *Hae*III site. (A) Primers that discriminate between the a-type and t-type arrays differ at the 3'-nucleotide and carry a common 5'-TAG sequence (underscored). (B) MVR principle showing the products of a-type and t-type amplification reactions using low concentrations of the 32-TAG-A and 32-TAG-T primers in parallel reactions, under which conditions there is expected to be a single priming event from each minisatellite molecule. These initial products carry sites for the 32D and the TAG primers, which are present at high concentrations and can therefore efficiently amplify the initial products to generate a set of products spanning between the 32D priming site and each a-type or t-type repeat unit in the minisatellite. PCR products are separated through an agarose gel and detected by Southern blot hybridization.

(24) and so only some guidelines on how this technique is performed will be considered.

Normally, isolated human DNA is used directly for the MVR-PCR amplification. However, if new allelic mutations are to be detected, digested human DNA should be size fractionated by agarose gel electrophoresis prior to MVR-PCR. This ensures that a series of DNA fractions are collected that contain DNA fragments that are smaller than the progenitor alleles and therefore should contain *de novo* deletions. Without size fractionation the detection of new mutations can be very difficult since contamination from the progenitor alleles, due to PCR amplification, can mask the detection of 'new' mutations.

Standard MVR-PCR amplifications should be performed using between 0.3 and 0.5 µg of human DNA in a final volume of 50 µl. The amplification reaction should contain all standard components for a PCR although the amplification cycles vary slightly from 'normal' PCR amplifications. For example, for MS32 allele mapping there is an extension time of 4 min which is performed for between 25 and 30 cycles. Following the PCR, the amplified alleles are detected on a 0.8–1% agarose gel and the separated products visualized by ethidium bromide staining. If single molecular analysis is performed, the PCR should be carried out in one-fifth of the volume of a standard amplification reaction and normally the detection method of choice is Southern blot hybridization using, in this case, an MS32 radiolabeled minisatellite probe.

11.14 Microsatellites

One limitation of VNTRs (Section 11.12) is that they are often located near telomeres. Another form of polymorphic repeat that does not show this bias is microsatellites, also known as TG or CA repeats. There are some 50 000 to 100 000 such repeats each of 10–60 copies of the dinucleotide. These repeats also show variation in length and therefore provide useful molecular markers (25). PCR primers can be defined from clones carrying microsatellites that allow the amplification of the dinucleotide repeat unit, but avoiding any complementarity to *Alu* repeat sequence. Since the variability in length is due to increases or decreases in the number of dinucleotides, it is important that the fragments being amplified are relatively small, consisting mainly of the repeat region, so that small dinucleotide differences can be readily resolved. There is a common problem observed with microsatellite analysis due to extra or so-called ghost bands. These may arise due to mispairing slippage of the DNA polymerase during copying of the dinucleotide repeat units.

The products of PCR can be detected either by $[\gamma^{32}P]$ end-labeling one of the primers, or incorporating $[\alpha^{32}P]$-dCTP into the PCR product, with subsequent detection by autoradiography or a phosphorimager. Increasingly, fluorescent end-labeled primers are being used with automated sequencer detection. Fluorescent labels such as FAM, HEX, ROX or TET (Chapters 3 and 9) can be used. However, one potential problem with this approach is the cost of primer synthesis as the fluorescent group additions can be expensive. This is not an issue for high-throughput large laboratories that will use all the primers, but for smaller laboratories perhaps only a small quantity of the labeled primer will be used and the remainder will be wasted. To overcome this potential problem a common fluorescently labeled primer should be utilized that can be used in each analysis, together with sequence-specific primers for the microsatellites. In an example given by Schuelke (26) the M13 (–20) primer (Chapter 5) was selected. The basic strategy requires that the forward sequence-specific primer is synthesized with a 5'-extension corresponding to the commonly used sequence and the reverse primer is designed as a normal primer with no extension. These unlabeled primers are inexpensive. The PCR is set up with the unlabeled forward primer (2 pmol), the labeled common primer (8 pmol) and the reverse primer (8 pmol). There is a lower concentration of the forward primer than reverse and the PCR is conducted initially at a high annealing temperature. As the reaction proceeds the forward primer leads to incorporation of the common primer tail into the accumulating products and also becomes depleted. During the later stages of the PCR the annealing temperature is reduced to allow the labeled common primer to take over the role of forward primer in the PCR, thereby labeling the products for subsequent detection. A similar approach could be adopted for other forms of fluorescent-labeling experiments.

11.15 Sensitive PCR for environmental and diagnostic applications

PCR provides an important tool for the sensitive detection of DNA as an indicator of the presence of an organism at low levels. Examples include:

- the detection of genetically modified organisms within a background of nonmodified material, for food, labeling or regulatory purposes. An example would be the detection of transgenic soy or maize within a bulk batch of predominantly nontransgenic material;
- the rapid detection of pathogenic microorganisms that either are present at very low levels and would require significant time and effort to culture, or are not culturable.

In detection of GM soy or maize it has proved possible to use primers for either single PCRs or multiplex PCRs that contain an internal positive control (27). Detection of contaminating GM maize was possible down to 0.001% dry weight and GM soy bean to 0.01% dry weight. The multiplex procedure reduced the detection sensitivity by 10-fold.

Organisms that cannot be cultured, or that require exotic and time-consuming conditions, are ideal candidates for sensitive PCR assays. However, the use of PCR is sometimes limited by the amount of DNA sequence information that is known from the organism concerned. The amplified sequences must be specific to the target organism, particularly as there are likely to be significant amounts of nontarget DNA present; for example, host DNA in detection of an infective pathogen.

A variety of diagnostic kits are available for common human viral pathogens, including human immunodeficiency virus, human T-cell lymphotrophic virus and hepatitis B virus, and the number and diversity of these and their routine use will also undoubtedly increase. Kits are now available for identification of a range of bacterial pathogens. For example the MicroSeq™ 16S kit (PE Biosystems) is based on PCR amplification of the 16S ribosomal DNA with identification by DNA sequence analysis. Two sequencing formats exist, either complete gene sequencing or, more rapidly, the first 500 nucleotides of the gene which is sufficient for identifying many bacteria, though is not as comprehensive as the complete gene analysis. In addition there is a database of more than 1200 bacterial 16S rRNA sequences with which to compare the data generated to facilitate accurate identification. Taqman® Pathogen Detection Kits (PE Biosystems) are also available, for detection of food-borne contaminants such as *Escherichia coli* O157 and some 60 other serotypes. The system is based on automated detection and the provision of a Yes/No answer rather than complex manual interpretation of amplification data.

A range of kits in both manual and fully automated formats are available from Roche Diagnostic Systems. These allow highly sensitive detection of infectious organisms before they become symptomatic or elicit other responses such as antibody accumulation. The AMPLICOR™ range of products detect HIV, hepatitis C, *Chlamydia trachomatis*, *Neisseria gonorrhoeae* and *Mycobacterium tuberculosis* and the last four are also available for full automation using the COBAS AMPLICOR™ instrument that combines thermal cycling, automated pipetting, incubation, washing and reader modules for diagnostic analysis.

Automation and simplification of procedures, including the avoidance of cross-contamination and simple detection, is important for ensuring wider application of PCR-based tests. An interesting innovation from Johnson and Johnson Clinical Diagnostics is based on a sealed blister system for complete automation of the PCR and subsequent product detection (28). A series of blisters contain the reagents for PCR, capture of products, washing and detection of product. The sample is introduced into the reaction pouch that contains biotinylated primers and reagents for hot-start PCR. After PCR, the final denatured products are physically forced into the next blister for streptavidin–horseradish peroxidase (HRP) capture of products, then a wash solution. Product detection is by immobilized oligonucleotide probes specific for target organisms, and detection relies upon formation of a colored product by the action of the HRP bound to the oligonucleotide probe via the specific PCR product.

Environmental testing for bacterial contamination in water supplies or soil can be important for disease prevention, and PCR amplification coupled with hybridization detection provides a highly sensitive tool for detection of levels as low as 1 cell per ml water. For example *Legionella pneumophila*, which is responsible for most cases of Legionnaire's disease, is notoriously difficult to culture. The MicroSeq™ 16S kit (PE Biosystems) discussed above provides a useful tool for such analyses. A recent report illustrates the use of a PCR/hybridization strategy to discriminate efficiently between *E. coli* and closely related bacteria by PCR analysis of DNA extracts from soil (29). The amplification target was the 16S ribosomal DNA sequence, and high-stringency annealing conditions of 72°C, well above the apparent T_m of the primers, was required for specificity. Other reaction conditions such as $MgCl_2$ concentration and times were empirically optimized. The use of Southern blot analysis of products allowed detection at levels as low as 10 genome equivalents.

11.16 Screening transgenics

An obvious use of PCR is for screening genomes for the presence of a particular gene as part of transgenic organism research programs, whether bacterial, fungal, plant or animal. If a construct has been introduced into an organism then usually some selectable marker is also introduced, allowing the initial selection of the transgenic lines. However, stable inheritance of the transgene is not guaranteed and it is usually necessary to analyze the genomic DNA to ensure all aspects of the transgene are present. Preliminary analysis of this type can be performed by PCR using primers designed from different elements of the introduced DNA. For example, one primer may correspond to a sequence in the upstream promoter region while another may be within the coding region (*Figure 11.7*). Since the DNA that has been introduced is well characterized it is possible to accurately predict the sizes of fragments that should be generated. Southern blot analysis can be performed to confirm the identity of the products. Ultimately it is safest to sequence the PCR fragments that have been generated to ensure that the DNA sequence introduced has not been altered. Unfortunately this level of analysis is too often neglected and fragment sizes alone are often taken as indicators of appropriate integration. Clearly if required the regions flanking the transgene insertion site can be isolated by using methods described in preceding sections as the situation is analogous to a transposon insertion line.

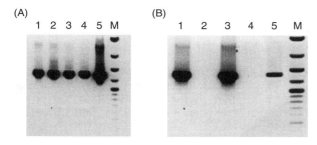

Figure 11.7

PCR screening of genomic DNA from transgenic plant tissue for the presence of intact transgenes. (A) PCR of samples 1–5 using gene-specific primers shows the presence of the target gene in all the transgenic samples. (B) PCR of samples 1–5 with a gene-specific and a promoter-specific primer demonstrates that only samples 1, 3 and 5 carry an intact version of the transgene. Presumably some rearrangement has led to the loss of part of the constructs in transgenic samples 2 and 4. Photograph kindly provided by A. Neelam (University of Leeds).

Further reading

Reeder DJ (1999) Impact of DNA typing on standards and practice in the forensic community. *Arch Path Lab Med* **123**: 1063–1065.

Savelkoul PH, Aarts HJ, de Haas J, Dijkshoorn L, Duim B, Otsen M, Rademaker JL, Schouls L, Lenstra JA (1999) Amplified-fragment length polymorphism analysis: the state of an art. *J Clin Microbiol* **37**: 3083–3091.

Vos P, Kuiper M (1998) AFLP analysis. In Caxtons-Arvolles G, Gresshoff PH (eds) *DNA Markers, Protocols, Applications and Overviews*, pp. 115–131. John Wiley & Sons, New York.

References

1. Sanger F, Coulson AR, Friedmann T, Air GM, Barrell BG, Brown NL, Fiddes JC, Hutchison CA, Slocombe PM, Smith M (1978) The nucleotide sequence of bacteriophage phiX174. *J Mol Biol* **125**: 225–246.
2. Fleischmann RD, Adams MD, White O, Clayton RA, Kirkness EF, Kerlavage AR, Bult CJ, Tomb JF, Dougherty BA, Merrick JM (1995) Whole-genome random sequencing and assembly of *Haemophilus influenzae* Rd. *Science* **269**: 496–512.
3. Goffeau A, Barrell BG, Bussey H, Davis RW, Dujon B, Feldmann H, Galibert F, Hoheisel JD, Jacq C, Johnston M, Louis EJ, Mewes HW, Murakami Y, Philippsen P, Tettelin H, Oliver SG (1996) Life with 6000 genes. *Science* **274**: 546.
4. Mewes HW, Albermann K, Bahr M, Frishman D, Gleissner A, Hani J, Heumann K, Kleine K, Maierl A, Oliver SG, Pfeiffer F, Zollner A (1997) Overview of the yeast genome. *Nature* **387**: 7–8, Suppl.
5. Lashkari DA, DeRisi JL, McCusker JH, Namath AF, Gentile C, Hwang SY, Brown PO, Davis RW (1997) Yeast microarrays for genome wide parallel genetic and gene expression analysis. *Proc Natl Acad Sci USA* **94**: 13057–13062.
6. Wodicka L, Dong HL, Mittmann M, Ho MH, Lockhart DJ (1997) Genome-wide expression monitoring in *Saccharomyces cerevisiae*. *Nature Biotechnol* **15**: 1359–1367.
7. Donis-Keller H, Green P, Helms C, Cartinhour S, Weiffenbach B, Stephens K, Keith TP, Bowden DW, Smith DR, Lander ES (1987) A genetic linkage map of the human genome. *Cell* **51**: 319–337.
8. Myers RM, Lumelsky N, Lerman LS, Maniatis T (1985) Detection of single base substitutions in total genomic DNA. *Nature* **313**: 495–498.
9. Orita M, Iwahana H, Kanazawa H, Hayashi K, Sekiya T (1989) Detection of polymorphisms of human DNA by gel electrophoresis as single-strand conformation polymorphisms. *Proc Natl Acad Sci USA* **86**: 2766–2770.
10. Zhu XQ, Gasser RB (1998) Single-strand conformation polymorphism (SSCP)-based mutation scanning approaches to fingerprint sequence variation in ribosomal DNA of ascaridoid nematodes. *Electrophoresis* **19**: 1366–1373.
11. Xiao W, Oefner PJ (2001) Denaturing High-Performance Liquid Chromatography: A review. *Hum Mut* **17**: 439–474.
12. Barany F (1991) Genetic disease detection and DNA amplification using cloned thermostable ligase. *Proc Natl Acad Sci USA* **88**: 189–193.
13. Newton CR, Graham A, Heptinstall LE, Powell SJ, Summers C, Kalsheker N, Smith JC, Markham AF (1989) Analysis of any point mutation in DNA – the amplification refractory mutation system (ARMS). *Nucleic Acids Res* **17**: 2503–2516.
14. Konieczny A, Ausubel FM (1993) A procedure for mapping *Arabidopsis* mutations using co-dominant ecotype-specific markers. *Plant J* **4**: 403–410.
15. Michaels SD, Amasino RM (1998) A robust method for detecting single-nucleotide changes as polymorphic markers by PCR. *Plant J* **14**: 381–385.
16. Neff MM, Neff JD, Chory J, Pepper AE (1998) dCAPS, a simple technique for the genetic analysis of single nucleotide polymorphisms: experimental applications in *Arabidopsis thaliana* genetics. *Plant J* **14**: 387–392.
17. Jordan B, Charest A, Dowd JF, Blumenstiel JP, Yeh RF, Osman A, Housman DE, Landers JE (2002) Genome complexity reduction for SNP genotyping analysis. *Proc Natl Acad Sci USA* **99**: 2942–2947.
18. Vos P, Hogers R, Bleeker M, Reijans M, van de Lee T, Hornes M, Frijters A, Pot J, Peleman J, Kuiper M, Zabeau M (1995) AFLP – a new technique for DNA fingerprinting. *Nucleic Acids Res* **23**: 4407–4414.
19. Bachem CWB, Hoeven van der RS, Bruijn de SM, Vreugdenhil D, Zabeau M (1996) Visualization of differential gene expression using a novel method of RNA fingerprinting based on AFLP: analysis of gene expression during potato tuber development. *Plant J* **9**: 745–753.

20. Houck CM, Rinehart FP, Schmid CW (1979) A ubiquitous family of repeated DNA sequences in the human genome. *J Mol Biol* **132**: 289–306.
21. Szmulewicz MN, Novick GE, Herrera RJ (1998) Effects of *Alu* insertions on gene function. *Electrophoresis* **19**: 1260–1264.
22. Anderson TD, Ross JP, Roby RK, Lee DA, Holland MM (1999) A validation study for the extraction and analysis of DNA from human nail material and its application to forensic casework. *J Forensic Sci* **44**: 1053–1056.
23. Jeffreys AJ, MacLeod A, Tamaki K, Neil DL, Monckton DG (1991) Minisatellite repeat coding as a digital approach to DNA typing. *Nature* **354**: 204–209.
24. Jeffreys AJ, Neumann R, Wilson V (1990) Repeat unit sequence variation in minisatellites: a novel source of DNA polymorphism for studying allelic variation and mutation by single molecule analysis. *Cell* **60**: 473–485.
25. Litt M, Luty JA (1989) A hypervariable microsatellite revealed by *in vitro* amplification of a dinucleotide repeat within the cardiac-muscle actin gene. *Am J Hum Genet* **44**: 397–401.
26. Schuelke M (2000) An economic method for the fluorescent labeling of PCR fragments. *Nature Biotechnol* **18**: 233–234.
27. Hurst CD, Knight A, Bruce JJ (1999) PCR detection of genetically modified soya and maize in foodstuffs. *Mol Breeding* **5**: 579–586.
28. Findlay JB, Atwood SM, Bergermeyer L, Chemelli J, Christy K, Cummins T, Donish W, Ekeze T, Falvo J, Patterson D, Puskas J, Quenin J, Shah J, Sharkey D, Sutherland JWH, Sutton R, Warren H, Wellman J (1993) Automated closed-vessel system for *in vitro* diagnostics based on polymerase chain reaction. *Clin Chem* **39**: 1927–1933.
29. Sabat G, Rose P, Hickey WJ, Harkin JM (2000) Selective and sensitive method for PCR amplification of *Escherichia coli* 16S rDNA genes in soil. *Appl Environ Microbiol* **66**: 844–849.

Index